ANIMAL KINGDOM

ANIMAL KINGDOM

By

Dr. DIGUMARTI BHASKARA RAO
Reader and Research Director
R.V.R. College of Education
D-43 (277) S.V.N. Colony
Guntur 522 006
(India)

DISCOVERY PUBLISHING HOUSE PVT. LTD.
NEW DELHI-110 002

First Published-1995
Reprinted-2011

ISBN 81-7141-274-2

Published by
DISCOVERY PUBLISHING HOUSE PVT. LTD.
4831/24, Ansari Road, Prahlad Street,
Darya Ganj, New Delhi-110002 (India)
Phone: 23279245 • Fax: 91-11-23253475
E-mail: dphbooks@rediffmail.com
dphtemp@indiatimes.com

Printed at:
Mehra Offset Press
Delhi

PREFACE

There are over a million described animal species. Of this number, about 95 per cent are invertebrates and the remaining are vertebrates.

The animal kingdom is generally believed to have originated in Archeozoic Oceans long before the first fossil record. Every major phylum has some marine representatives or groups. From the ancestral marine environment, different groups of animals have invaded fresh water, and some have moved onto land.

After settling well, the man has developed interest in studying living organisms, which is termed as Biology. Zoology is a branch of Biology that deals with animal life and Botany is another branch concerned with plant life.

Aristotle, 384–322 B.C., laid the foundation of Zoology. He studied the animals for over a decade and applied a crude method of classification to classify them

John Ray, 1627–1705 A.D., made the earliest attempt to classify the animals systematically. He used the term genera for the groups containing closely related species, and also developed a key for the identification of a given animal.

Karl van Linnae, 1707–1795, was the real founder of taxonomy who published the "Systema Naturae" in 1735. He mentioned in it the binomial names and the method of classification. He formulated six classes—Quadrupods, Birds, Reptiles, Fish, Insects and Worms—of animals. He divided each class into orders, each order into genera, each genera into species on the basis of some important differences. Each species was given a double name, made up of the name of the genus and of the species itself—which is known as binomial nomenclature. He even changed his name to Carolus Linnaeus on the basis of his binomial nomenclature.

Georges Leopold Cuvier, 1769–1832 A.D., in 1800 added another category, the phylum, which includes all the classes of animals with common body structure.

Ernst Haeckel, in 1864, and E. Ray Lankester, in 1877, outlined the principal features of the zoological classification which are in use to date.

The entire animal world, according to natural classification, is subdivided into successively smaller and more restricted groups, ending finally with the species, the only real zoological entity.

The animals have been classified in various ways by different scientists. The broad classification of animal kingdom given in this book is the one prepared by L.H. Hyman in 1940 with a few changes. All the major phyla have been explained eliminating the minor phyla which have limited importance in general science education.

The general characters of each phylum or class will develop a broader outlook towards it. The classification gives a clear picture in identifying a species. Each category in classification is supported by suitable illustrations.

As the book puts all the information about classification of animal kingdom at one place, the readers can use this book as a ready reckoner to understand animal classification. Constructive criticism is hereby invited with great pleasure.

Dr. D. Bhaskara Rao

To

Mr. Lokeswara Rao

Mrs. Naga Ratnam

Mr. Satya Narayana

Mrs. Vimala Kumari

Mr. Venkateswara Rao

Mrs. Vijaya Lakshmi

Mr. Venkata Rao

Mrs. Prameela

CONTENTS

INTRODUCTION

Life according to an estimate has originated in this world at about 3,500 million years ago. It is difficult to define life. However, life can be understood by the features such as growth, locomotion, metabolism, irritability, reproduction, etc., exhibited by the living organism. These characters help to distinguish living things from the non-living.

The term Biology was coined by Lamarck and Treviranus. The science dealing with study of life is called Biology (Gr.bios = Life; Logos = study). Living things are grouped into animals and plants. The science dealing with the animal life is called Zoology (Gr. Zoon= animal; Logos= study), whereas, the science dealing with the plant life is called Botany (Gr. botane= herb).

The term Zoology includes various aspects of the study of animals. To understand these aspects, a systematic approach is needed. Hence, Zoology is studied under various branches.

BRANCHES OF ZOOLOGY

1. **Morphology** (Gr. morpho = form+logos=study): It is the study of the structure of animals. Morphology is distinguished into external morphology and internal morphology depending on the study of external and internal characters respectively.

2. **Physiology** (Gr. physis = nature+logos = study). It is the study concerned with the function of various organs and systems. This is further divided into:

a. **Cellular physiology**: This is the study of the functions of cell and its constituents.

b. **Comparative physiology:** The study concerned with the comparative account of functional mechanisms in different animals.

c. **Physiological chemistry (Biochemistry):** The study concerned with the animal functions involving chemical reactions.

d. **Systematic physiology:** The study concerned with the function of organs and organ systems.

3. **Cytology** (Gr. Kyotos = hollow vessel+ logos = study) : The study of the structure and functions of cells and their constituents.

4. **Embryology** (Gr. embryon =embryo + logos = study): It is the study of the development of animals form the unicellular zygote to multicellular adult.

5. **Ecology** (Gr. oikos = house + logos = study): The study of the relationship between organisms and environment.

6. **Palaeontology** (Gr. palaeos = ancient + onto = existing + logos = study): It is the study of fossils. Fossils are the body remains of the animals and plants of the remote past.

7. **Evolution** (L, e= out + volvere= develop) : The study of the unfolding of life formed due to descent with modification.

8. **Genetics** (Gr. genesis - origin) : It is the study of heredity and variations.

9. **Histology** (Gr. histos = tissue + logos = study): It is the study of structure and functions of tissues.

10. **Zoogeography** (Gr. zoon= animal + ge = earth + graphein = to write): It is the study of the distribution of animals all over the earth.

11. **Taxonomy** (Gr. taxis = organization + nomos = law): It is the study of the classification of animals and plants.

12. **Anatomy** (Gr. ana = up + tome = cutting): It is the study of the internal structures of organisms.

SPECIALIZATIONS

No.	Specialization	Study
1.	**Anthropology**	***Study of apes and man***
2.	**Arthrology**	***Study of joints***
3.	**Batracology**	***Study of frogs***
4.	**Carcinology**	***Study of crustaceans***
5.	**Conchology**	***Study of molluscan shells***
6.	**Dermatology**	***Study of skin***
7.	**Entomology**	***Study of insects***
8.	**Helminthology**	***Study of helminth worms***
9.	**Haemotology**	***Study of blood***
10.	**Herpetology**	***Study of Reptilia and Amphibia***
11.	**Ichthyology**	***Study of fishes***
12.	**Mammology**	***Study of mammals***
13.	**Mastology**	***Study of breast***
14.	**Myology**	***Study of muscles***
15.	**Malacology**	***Study of molluscs***
16.	**Nephrology**	***Study of kidney***
17.	**Neurology**	***Study of nervous system***
18.	**Ophiology**	***Study of snakes***
19.	**Ornithology**	***Study of birds***
20.	**Osteology**	***Study of skeleton***
21.	**Odontology**	***Study of teeth***
22.	**Otology**	***Study of ears***
23.	**Pleurology**	***Study of lungs***
24.	**Phrenology**	***Study of brain***
25.	**Rhinology**	***Study of nose***
26.	**Saurology**	***Study of lizards***
27.	**Tricology**	***Study of hair***
28.	**Bionomics**	***Study of the relation of the organisms and its environment.***
29.	**Chronobiology**	***Study of duration of life***
30.	**Cytogenetics**	***Study of heredity from the point of view of cytology and genetics***

31.	**Epidemiology**	***Branch of medicine dealing with epidemic diseases***
32.	**Ethnology**	***Origin, distribution and distinguishing characteristics of race of mankind***
33.	**Ethology**	***Study of animal behaviour***
34.	**Gerontology**	***Study of old age, its phenomenon and diseases***
35.	**Hygiene**	***Science of health and its preservation***
36.	**Microbiology**	***Study of minute living organisms***
37.	**Mycology**	***Study of nervous system, its functions and disorders***
38.	**Neuropathology**	***Diseases of nervous system***
39.	**Orthopedics**	***Prevention, diagnosis and treatment of diseases and abnormalities of musculoskeletal system***
40.	**Phthisiology**	***Scientific study of tuberculosis***
41.	**Radiobiology**	***Effects of radiation on living organisms***
42.	**Toxicology**	***Study of poisons***
43.	**Notobiology**	***Study of germ free organisms***
44.	**Teratology**	***Study of abnormalities during the development of embryo***

TAXONOMY

Taxonomy is a branch of Biology, which deals with the study of identification, naming and classification of living things. ***Naming of plants and animals is called nomenclăture and orderly grouping of the plants and animals on the basis of their relationship and evolution is known as classification.***

There are about 1.5 million species of animals which have been named and described and the number is increasing day by day. The vast number of animals are known for their diversity of form, structure, habitat, habit, behaviour, etc. It would be impossible to study the animal world without a definite system of classification.

Aristotle was the first person to indicate how animals can be grouped. However, his classification is based on the habits and habitats of animals which cannot bring out evolutionary relationship between the animals. Hence, it is said to be artificial.

John Roy was the first person to attempt a systematic classification based on anatomical facts. **John Ray** developed the concept of species and genera. **Karl von Linnaeus**, a Swedish naturalist is the founder of Modern Taxonomy. He published a book entitled **Systema Naturae** in 1735, in which he mentioned the binomial nomenclature and classification of all the living things known to him. ***Later, Haeckel and Lankster (1864 & 1867) outlined the principles of classification which are followed even today.***

SPECIES

It is the basic unit of systematic classification of plants and animals. **Mayor** ***defined a species as a group of similar individuals which can freely interbreed among themselves, but reproductively isolated from such other groups***. It has distinct characters of its own. A species, often consists of sub-species or geographical races. A subspecies is defined as a geographically isolated subdivision of species, which differs genetically and taxonomically from the other subspecies.

NOMENCLATURE

The naming of plants and animals on the scientific basis is called **nomenclature**. The common names of plants and animals lead to confusion. Often, the same animal is known by different names in different parts of the world or in different regions of the same country. Further, the same common name may be used for different animals. Hence, there is a need for a scientific name which is to be used all over the world for the same species. The scientific names are the language of taxonomists.

BINOMIAL NOMENCLATURE

It was first introduced by **Linnaeus**. It is followed till to date all over the world. ***According to this system of nomenclature, scientifically every species of plant or animal has two names. The first name denotes the genus while the second denotes the species.*** Both the names are to be latinised and italicized. As a rule, the generic name is to be placed before the species name. The generic name is always a noun and begins with a capital letter. The species name is mostly an adjective and begins with a small letter.

Example

	Common name	Technical name
1.	Tape worm.	*Taenia solium*
2.	House fly	*Musca domestica*
3.	Honey bee	*Apis indica*
4.	Apple snail	*Pila virens*

TRINOMIAL NOMENCLATURE

It includes genus, species and subspecies. The subspecies name is also a Latin or latinized word and follows the species.

Example

I. The scientific name of a water flea (Cladoceran) is ***Moina micrura.***

There are 3 sub-species in it. They are:

1. ***Moina micrura micrura***
2. ***Moina micrura ciliata***
3. ***Moina micrura dubia.***

II. The scientific name of house crow is *Corvus splendens.* There are three subspecies in it.

Indian house crow	*Corvus splendens splendens*
Burmese house crow	*Corvus splendens insolens*
Ceylonese house crow	*Corvus splendens protegatus*

Animal taxonomists all over the world follow the rules framed by **International Commission on Zoological Nomenclature** in 1898.

TAUTONYM

It is the nomenclature in which the generic name and species name are one and the same.

Examples

Common name	Scientific name
Rotifier	*Rotaria rotaria*
Spotted deer	*Axis axis*
Fowl	*Gallus gallus*
Indian cobra	*Naza naza*
Gorilla	*Gorilla gorilla*
Rat	*Rattus rattus*

NATURAL CLASSIFICATION

The resemblance and differences in structure between different types of animals form the basis for classification. Classification based on homologies of organs reveals that how the different kinds of animals are derived from their ancestors. Such a classification which reveals the relationship of different groups of animals is called *natural classification*.

Animal kingdom is subdivided into successively smaller groups ending with the individual.

Species : It is the smallest taxon or basic unit of classification.

Genus : A genus is a group of closely related species.

Family : A family is a group of closely related genera.

Order : An order is group of closely related families.

Class : A class is a group of closely related orders.

Phylum : A phylum is a group of closely related classes.

ANIMAL KINGDOM

All the phyla constitute the animal kingdom. Sometimes, intermediate taxa like, subfamilies, suborders, superorders, subclasses, superclasses, subphyla, subkingdoms, etc., are erected to show intimate relationship in the concerned groups. Thus, the animal kingdom is comparable to a big branched tree. The successively smaller taxa are comparable to the successively smaller branches of the tree.

Taxonomic Position of Cockroach

Kingdom	*Animalia*
Subkingdom	*Metazoa*
Branch	*Eumetazoa*
Grade	*Bilateria*
Division	*Proterostomia*
Subdivision	*Schizocoelomata*
Phylum	*Arthropoda*
Subphylum	*Mandibulata*
Class	*Insecta*
Subclass	*Pterygota*
Order	*Orthoptera*
Family	*Blattidae*
Genus	*Periplaneta*
Species	*Americana*

In addition to major phyla, a number of minor phyla like Rotifera, Ectoprocta, Phoronida, Brachiopoda, Chaetognatha are also present in the animal kingdom.

Animals have been classified in various ways by different authors. Earlier the classification is based on structural criteria like number of cells, germ layers, nature of body cavity, etc. At present, other criteria based on biochemical and physiological differences are also used to solve the riddles of classification. *The broad classification of the animal kingdom given here is the one proposed by L.H. Hyman in 1940 with a few changes.*

Kingdom Animalia comprises of all diversified animals of the Biosphere. It is divided into two subkingdoms.

1. Subkingdom PROTOZOA

It consists of all acellular animals. They are generally microscopic, either free living or parasitic. It includes the phylum Protozoa.

Eg: Amoeba, Euglena, Paramoecium are free living. *Plasmodium, Entamoeba histolytica* are parasites of man.

2. Subkingdom METAZOA

All the multicellular animals are included in it. It is divided into two branches.

I. PARAZOA It includes multicellular animals with numerous pores, canal system and without tissue grade. Only phylum Porifera is present in it. It consists of sponges which are sedentary aquatic animals with spongocoel lined by choanocytes and performed with numerous pores called ostia.

Eg: Leucosolenia, Euplectella, Euspongia, Cliona.

II. EUMETAZOA It consists of either radially or bilaterally symmetrical animals. It is subdivided into two grades namely, Radiata or Diploblastica and Bilateria or Triploblastica.

Radiata : It is the first formed metazoan group with tissue grade, radial symmetry and diploblastic body wall. It consists of the phylum Cnidaria.

Eg: Hydra, Obelia, Aurelia, Sea anemone.

Bilateria : It comprises of metazoans with 3 germ layers in the body wall and bilateral symmetry. It is subdivided into two divisions namely, Protostomia and Deuterostomia.

A. ***Protostomia*** : This division consists of triplo blastic animals in which the mouth arises form or near the blastopore. It consists of 3 subdivisions.

a. Acoelomata : It includes protostomes without body cavity. The space between body wall and alimentary canal is filled with parenchyma. Phylum Platyhelminthes is included in it.

Eg: *Tapeworm, Liver fluke, Blood fluke, Dugesia.*

b. Pseudocoelomata : It consists of protostomes with a false body cavity. A body cavity which is not lined by coelomic epithelia or a ruminant of embryonic blastocoel is called pseudocoel or false cavity. This subdivision consists of Nemathelminthes.

Eg: *Ascaris, Hook worm, Wuchereria.*

c. Schizocelous Coelomata: These are the protostomes with schizocoel as body cavity. A cavity, formed by the splitting of embryonic mesoderm is called schizocoel. There are a number of minor and major phyla in it.

Examples

Phylum Annelida	-	*Earthworm, Leech, Neanthes.*
Phylum Arthropoda	-	*Cockroach, Prawn, Scorpion.*
Phylum Mollusca	-	*Pila, Unio, Sepia, Chiton.*

B. ***Deuterostomia*** : This division includes the triploblastic Metazoa in which the blastopore generally becomes an anus. There is only one subdivision that is, *Enterocoelous coelomata* in it. The body cavity is enterocoel. There are two major phyla besides some minor phyla in it. Major phyla are—

1. Phylum Echinodermata	-	*Star fish, Brittle star, Sea-lily*
2. Phylum Chordata	-	*Amphioxus, Ascidian, Fishes, Amphibians, Reptiles, Birds and Mammals.*

The bulk of the animal kingdom comprises of invertebrates, i.e., animals without backbone. The invertebrates account for 95% of the animals. The range is from microscopic protozoans to the largest living invertebrate, *Architeuthis*. They show a lot of diversity in structure and organization unlike the chordates which exhibit greater degree of similarity. Invertebrates like honey bees, silk worms, lac insects, prawns, lobsters, pearl oysters, earthworms, etc., are useful to mankind in several ways.

Parasitic invertebrates like sanguivorous insects, ship worms, termites, scorpions, insect pests, sea wasps, star fishes, etc., are harmful to mankind in different ways. Hence, a study of some of them would enable us to find out ways and means to destroy them, if they are harmful. Though, many of them appear to be neither harmful nor beneficial, each one of them plays its role in the maintenance of ecological balance.

There are over a million described species of animals. Of this number about 5% possess a backbone and are known as vertebrates. All others, comprising the greater part of the Animal Kingdom and are called invertebrates.

ANIMAL KINGDOM
Major Phyla

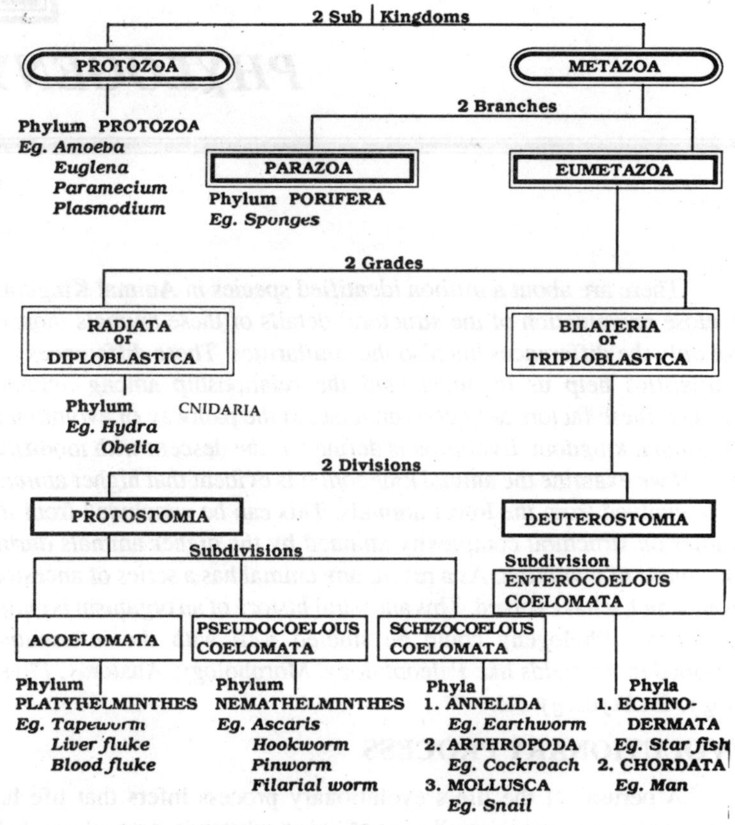

PHYLOGENY

There are about a million identified species in Animal Kingdom. A close examination of the structural details of these animals indicate not only the differences but also the similarities. These differences and similarities help us to understand the relationship among animals. Further, these factors help us to understand the pathway of evolution in the animal kingdom. Evolution is defined as the descent with modification. If we examine the animal kingdom it is evident that higher animals have evolved from the lower animals. This can be concluded from the studies on structural complexity attained by the higher animals during the course of evolution. As a result, any animal has a series of ancestors from which it has evolved. This ancestral history of an organism is called phylogeny. Phylogeny could be studied well with the information obtained in the fields like Paleontology, Morphology, Anatomy, Physiology, Embryology etc.

EVOLUTIONARY PROCESS

A perusal of the life's evolutionary process infers that life has originated in the sea as a small mass of living substance, protoplasm. Life has formed from the aminoacids and sugars which have undergone a series of chemical reactions forming nucleic acids. The formation of DNA, the self replicating molecule led to the formation of the life in the form of a primitive, simple single-celled organism. From this primitive cell, multicellular animals such as sponges have formed. These animals are not having tissue grade of organization. In due course, from such type of animals the multicellular diploblastic cnidarians have originated. These animals have shown the tissue grade of organization. The evolution of sponges and cnidarians from the single-celled organism can be evident from the embryological studies.

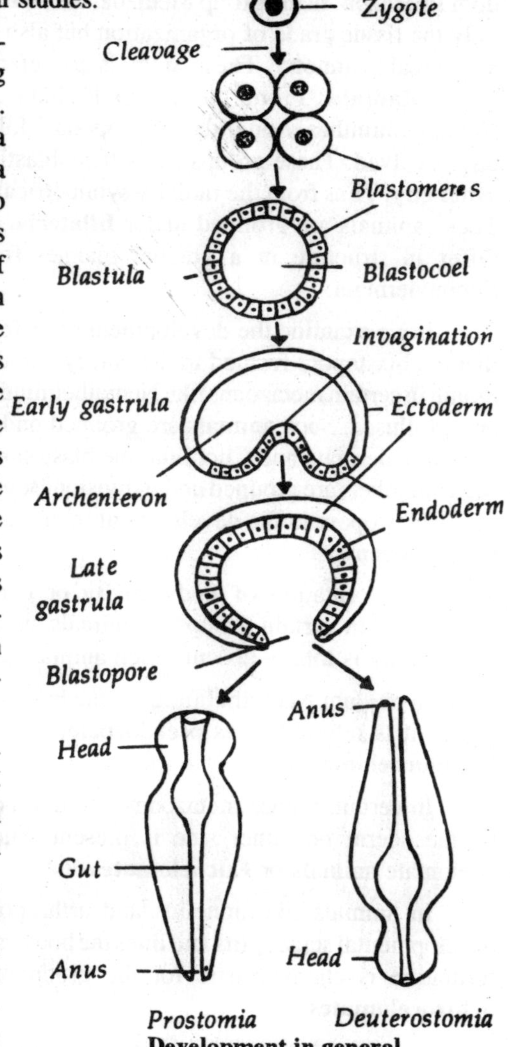

Development in general

In the development of an animal, egg represents the first stage. This egg develops into a single layered blastula and subsequently a two layered gastrula. In this developmental history of any animal, the egg can be comparable to the unicellular organisms like protozoans. The blastula represents the multicellular animals like sponges which are not having tissue grade of organization, whereas the gastrula represents the multicellular organisms (cnidarians) with tissue grade of organization.

If we examine different animals in Protozoa, Porifera and cnidaria, we can find single celled organisms, colonial organisms with a group of cells living together and colonial forms having flagellated collar-shaped cells (choanoflagellates), sponges and two layered animals like *Hydra*. All these animals indicate a progressive evolution from one stage to that of the other.

From the single celled stage certain organisms have evolved to multicellular organisms and formed the subkingdom **Metazoa** whereas the single celled organisms obtained as single cells forming the subkingdom **Protozoa**. In the group Metazoa, the cnidarians have developed not only the tissue grade of organization but also exhibit diploblastic body and radial symmetry. These animals are referred to as members of the group **Radiata.** From the group Radiata, the metazoans such as Nemathelminthes, Annelida, Arthropoda, Mollusca and Echinodermata have evolved. These groups show triploblastic body wall and bilateral symmetry. Thus from the radially symmetrical animals have originated. These animals are grouped under **Bilateria.** These bilaterians further differ in structure in a gradual manner from Nemathelminthes to Echinodermata.

If we examine the development of different metazoans, it is seen that the blastopore formed in the embryonic development becomes the mouth in certain metazoans like Nemathelminthes, Annelida, Arthropoda and Mollusca. Such animals are grouped under the term **Protostomia.** In Echinodermata and Chordata, the blastopore forms the anus. Hence, these members are grouped under division **Deuterostomia.** Thus The fate of the blastopore in the development of animals helps in understanding the phylogeny.

The formation of body cavity or coelom differs in different metazoans. In certain groups of animals such as Platyhelminthes, the body cavity is totally absent. Such animals are called **Acoelomates.**

In phylum Nemathelminthes, the body cavity is not a true cavity. Hence it is referred to as pseudocoelom and the members are called **Pseudocoelomates.**

In certain higher metazoans, the true body cavity which is lined by mesoderm on either side is present. Such forms are called true coelomate animals or **Eucoelomates.**

In animals like annelids and arthropods, the mesoderm in its developmental stage splits and lines the body cavity. Such type of coelom formation is characteristic for the group of animals referred to as **Schizocelomates.**

In Echinodermata and Chordata, the body cavity is formed from the mesodermal pouches formed as lateral evaginations of primary gut. These pouches enclose the coelom. The coelom is enclosed by coelomic epithilium. Such type of coelom formation is referred to as enterocoelom

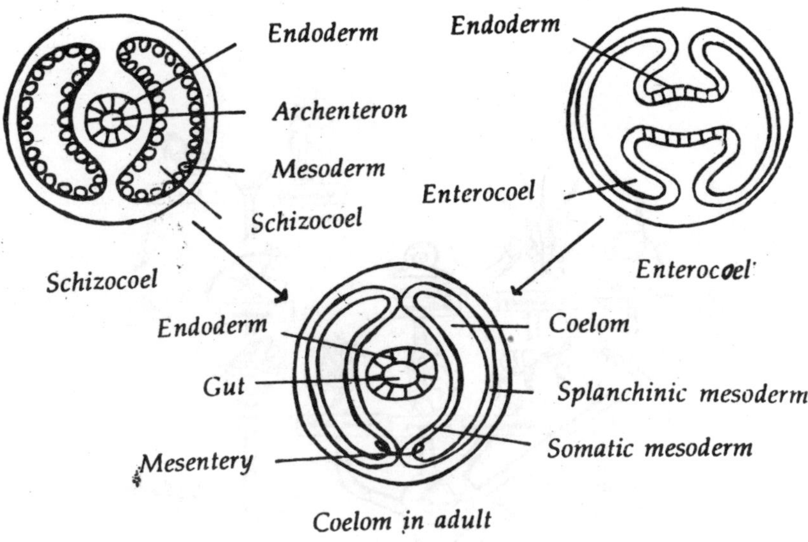

Formation of Coelom

type and the members exhibiting this type of coelom formation are called **Enterocoelomates.**

The evolutionary process in different animals can be followed by the following points.

1. Primitive protozoans led to the formation of colonial flagellates like *Volvox*. Such type of forms may have formed the primitive metazoans.
2. In coelenterates, the planula larva is seen in the developmental history. This is a diploblastic form. It may have settled on the bottom of the water bodies and started creeping or crawling on the bottom. This has changed the body shape and the form has become slightly flattened. Its symmetry has changed. Thus radially symmetrical planula larva has formed into bilaterally symmetrical metazoans.
3. The bilaterally symmetrical animals have formed a third layer of cells called mesoderm. This was led to the formation of triploblastic condition. Thus flatworms form the primitive triploblastic, bilaterally symmetrical animals.

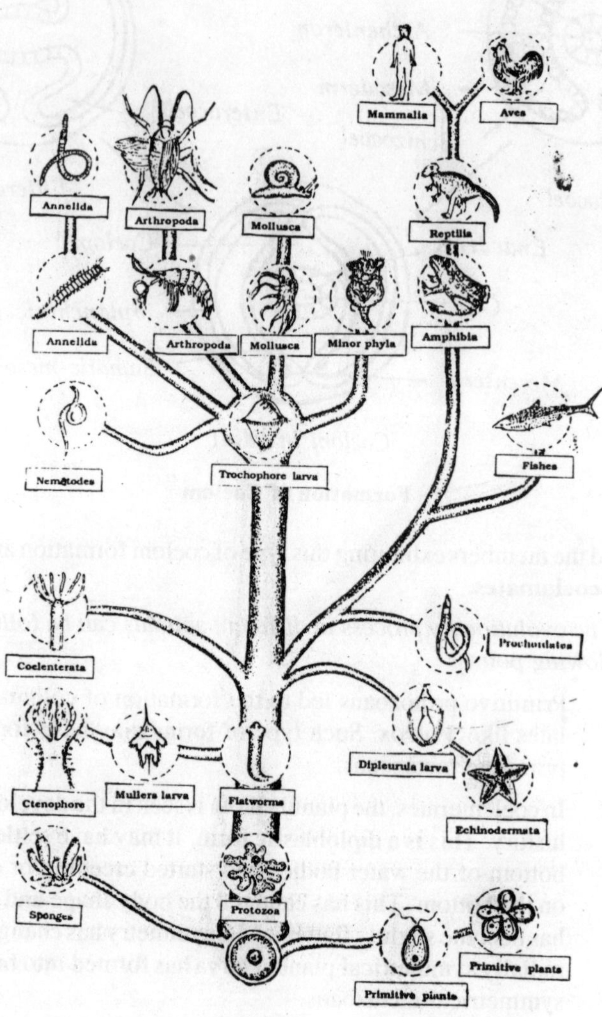

Phylogenetic Tree

4. The platyhelminthes have no body cavity. Hence, they are called acoelomates. From the acoelomates, pseudocoelomates such as round worms and eucoelomates such as arthropods, molluscs, echinoderms and chordates are formed.
5. In the eucoelomates, two groups such as schizocoelomates represented by arthropods and molluscs and enterocoelomates such as echinoderms and chordates have originated.
6. The presence of segmented body in Annelida and Arthropoda indicate affinity. Further, the presence of an annelid-like caterpillar seen in the life history of arthropods like butterfly suggest that arthropods have descended from annelid-like ancestors. Peripatus exhibits the characters of both annelids and arthropods. Thus, it stands as a connecting link between Annelida and Arthropoda suggesting that annelids give rise to arthropods.
7. The trocophore larva in Annelida resembles that of veliger larva in Mollusca indicating the affinity of annelids with molluscs.
8. **Marcus** studied the phylogeny of the animals and stated that flagellates gave origin to metazoans like poriferans and cnidarians. From these lower metazoans, schizocoelomates have originated on separate stalks. This is referred to as Marcus theory.
9. **Hanson** believes that the primitive protozoans hav evolved a poriferans on one side and true metazoans (acoelom flat-worms) on the other side. Coelenterates and flat worms have originated as separate groups from the acoelomates. The acoelomate flatworms gave rise to schizocoelomates on one side and enterocoelomates on the other side.
10. **Hyman** has proposed another theory on the evolution of animal kingdom. According to this theory, the primitive protozoans evolved into a colonial flagellates from which the sponges have originated on one side and a planula-like metazoan on the other side. The planula-like metazoan gave rise to coelenterates and acoelom flat worms as two separate groups. From the acoelom flatworms, the schizocoelomates and enterocoelomates are evolved as two separate stalks.

PROTOZOA

The Protozoa (Gr. protos = first + zoan = animal) are the smallest, simplest and the most primitive of all animals. Most of them are microscopic and they are not seen without the aid of a microscope. Hence, they were not known prior to the invention of the microscope. Even today, they are not known to layman. They are usually measured in microns. Each protozoan is represented by a single cell. Structurally, a protozoan is comparable with the cell of a multicellular metazoan, but functionally it is a complete organism.

Protozoa are often described as unicellular organisms, because all the general functions are performed by a single cell. But, it is preferable to call them non-cellular or acellular animals rather than unicellular. The specialized parts of a protozoan which perform various functions are called organelles (cell organs).

The Protozoa are cosmopolitan in distribution occurring all over the world. They live in aquatic habitats such as fresh, salt or brackish waters, in damp soil or decaying organic matter. Many are free-living and free-swimming whereas, others are sessile animals; certain protozoans live as parasites in other animals or plants, while a few lead symbiotic life. Most of the protozoans are solitary, while some are colonial. Many of them serve as food for other small animals.

The phylum name Protozoa was coined by Goldfuss in 1817. Protozoans were first of all observed by Antony van Leewenhock in 1671 and called them animalcules. The unicellular nature of the protozoans was first recognized by Von Siebold in 1845. About 50,000 species of

Protozoa have been described so far. The study of protozoans help in the improvement of our living standards.

GENERAL CHARACTERS

1. Protozoans are minute, usually microscopic and not visible without the aid of a microscope.
2. They are simplest and most primitive animals.
3. Protozoans are solitary or colonial. The individuals in colonies are alike and interdependent.
4. Unicellular or acellular or non-cellular animals containing one or more nuclei and organelles. There is no differentiation of cells, tissues and organs.
5. Body symmetry is bilateral, radial, spherical or none.
6. Body is naked or enveloped by a pellicle. Some animals are provided with shells.
7. Body form is constant or varied.
8. The single cell performs all the vital activities. There is no physiological division of labour.
9. Locomotion is effected by finger-like pseudopodia or whip-like flagella or hair-like cilia. Locomotory organelles are totally absent in some animals.
10. Nutrition is holozoic, holophytic or saprophylic.
11. Digestion is intracellular which occurs inside the food vacuoles.
12. Respiration takes place through general body surface by diffusion.
13. Excretion takes place through general surface of the body or through contractile vacuole. Contractile vacuole plays a role in osmoregulation.
14. Reproduction occurs by asexual and sexual methods. Asexual reproduction is by binary or multiple fissions and budding. Sexual reproduction is by fusion of gametes.
15. Life histories are complicated and alternation of generation is seen in some animals.
16. Encystment is common in many protozoans. Encystment helps to tide over the unfavourable conditions and in dispersal.
17. Mostly protozoans are free living, inhabiting fresh and marine waters and damp places while some are parasitic or symbiotic.

18. The body substance is not differentiated into somatoplasm and germplasm, hence they have no natural death (immortals).

CLASSIFICATION

The phylum Protozoa contains about 30,000 known species. It is divided into two sub-phyla and six classes. *The classification of Protozoa is based on the nature of locomotory organelles.*

SUB-PHYLUM I. PLASMODROMA

1. They are simple and primitive animals.
2. Locomotory organelles are pseudopodia or flagella. Locomotory organelles are absent in some.
3. The members show single or many nuclei but all are of the same kind.
4. Asexual reproduction is binary or multiple fissions.
5. Sexual reproduction is by the fusion of gametes.

The subphylum Plasmodroma includes four classes.

Class 1. Rhizopoda

1. Animals of this class are mostly amoeboid in shape.
2. Locomotion and ingestion of the food is by pseudopodia.
3. They are mostly free-living, but a few are parasitic.
4. Asexual reproduction is usually by binary fission.
5. Encystment is common.

Examples

Amoeba, Entamoeba, Elphidium, Actinophrys.

Class 2. Mastigophora

1. Body of the mastigophorans is covered by a thin pellicle.
2. Locomotion and food capture are by one or more whip- like flagella.
3. Both free living and parasitic animals constitute this class.
4. Asexual reproduction is mostly by longitudinal binary fission.
5. Encystment is common.

Examples

Euglena, Trypanosoma, Leishmania, Volvox, Giardia, Noctiluca, Proterospongia.

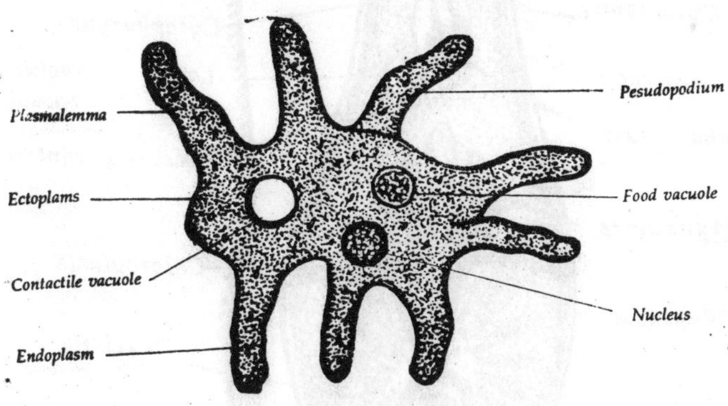

Amoeba

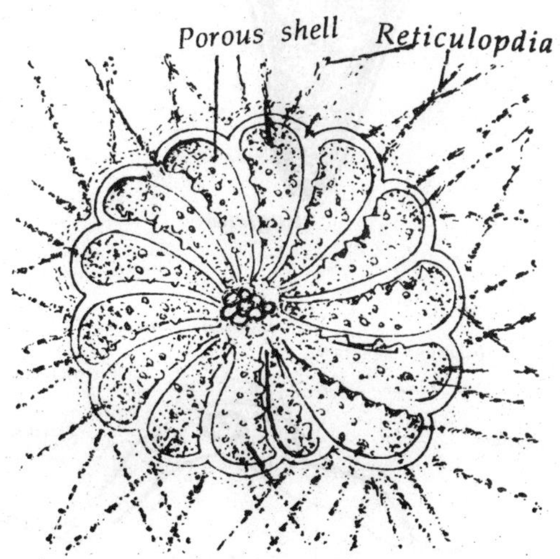

Elphidium

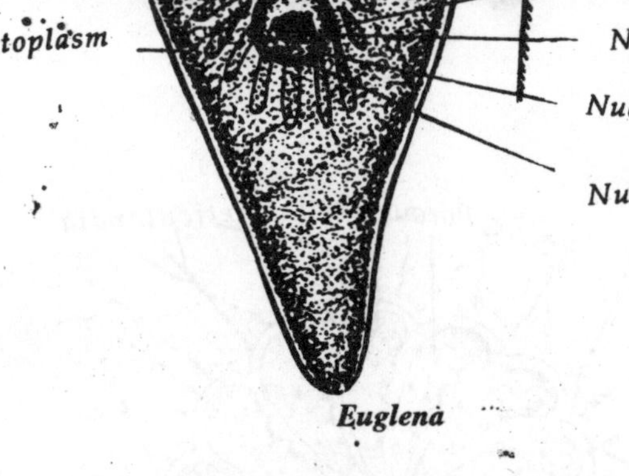

Euglena

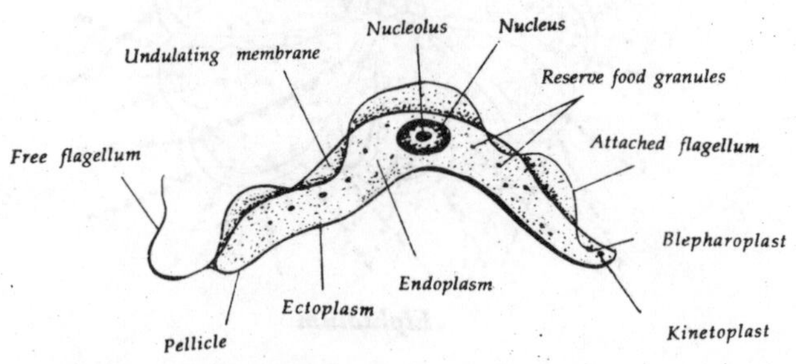

Trypanosoma Gambiense

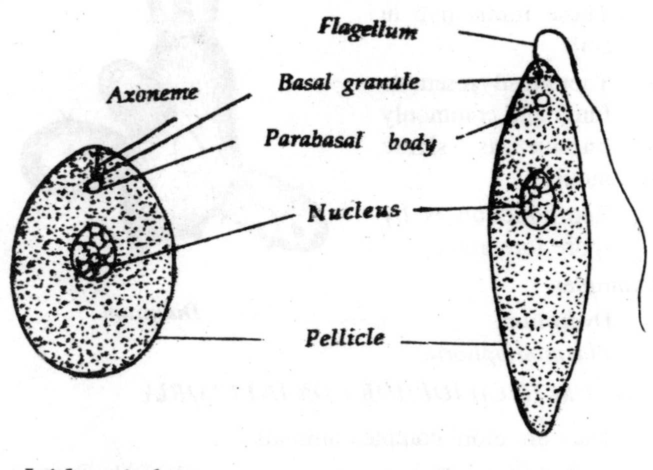

Leishmania form *Leptomonas form*

Class 3. Sporozoa

1. Spore formation is common.
2. They are parasitic animals without locomotory organelles and contractile vacuoles in the adult stage.
3. Reproduction is by asexual (multiple fission) and sexual methods.

Examples

Plasmodium, Monocystis, Eimeria, Sarcocystes, Babesia.

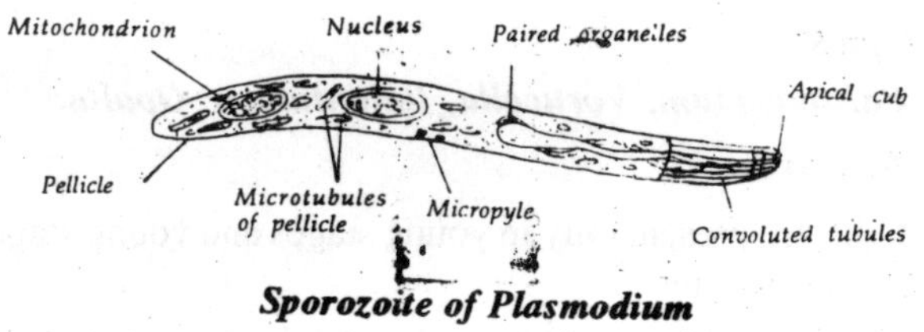

Sporozoite of Plasmodium

Class 4. Mycetozoa

1. These forms live in soil.
2. They mostly resemble fungi and commonly known as slime moulds.
3. Reproduction is by spore formation.

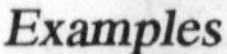

Examples

Dydinium, Plasmodiophora.

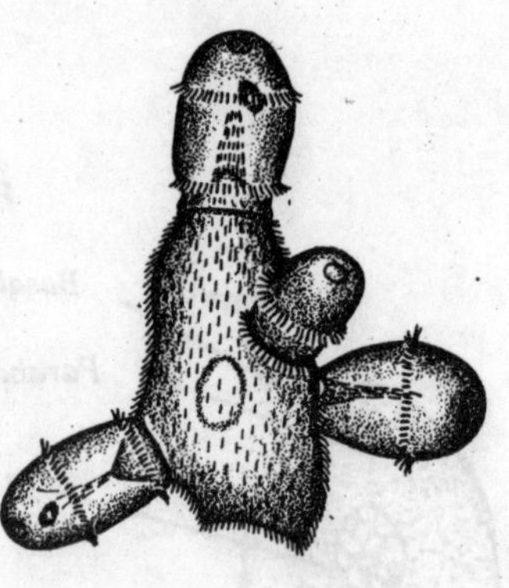

Didinium

SUB-PHYLUM II. CILIOPHORA OR INFUSORIA

1. They are more complex animals.
2. Locomotion is by cilia. Cilia may be replaced by the tentacles in adults.
3. Usually two types of nuclei are present. 4. Asexual reproduction is by binary fission or budding.
5. Sexual reproduction is by conjugation.

The sub-phylum Ciliophora includes two classes.

Class 1. Ciliata

1. Locomotion is by numerous hair-like cilia, present throughout life.
2. Asexual reproduction is by transverse binary fission.
3. Sexual reproduction is by conjugation, autogamy, endomixis etc.

Examples

Paramoecium, Vorticella, Balantidium, Opalina.

Class 2. Suctoria

1. Cilia are present only in young stages and young stages freely swim in water.
2. Tentacles are present in adults. Adults are sedentary, without cilia. Food capture and ingestion of food is by suctorial tentacles.
3. Asexual reproduction is mostly by budding.

Examples : ***Acineta, Ephelota.***

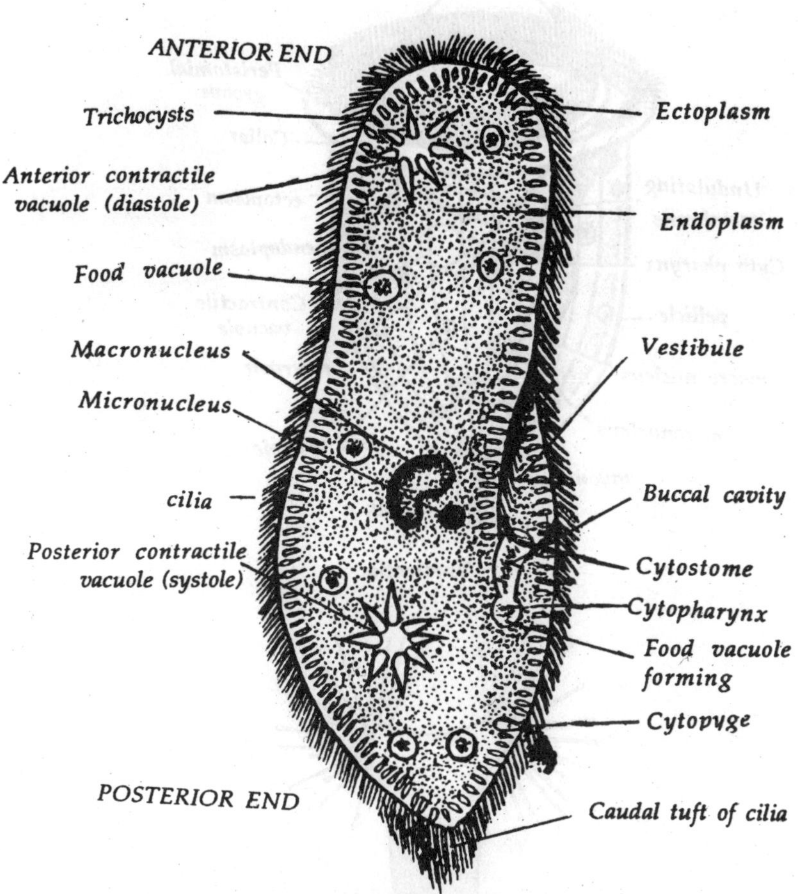

Paramoecium Caudatum

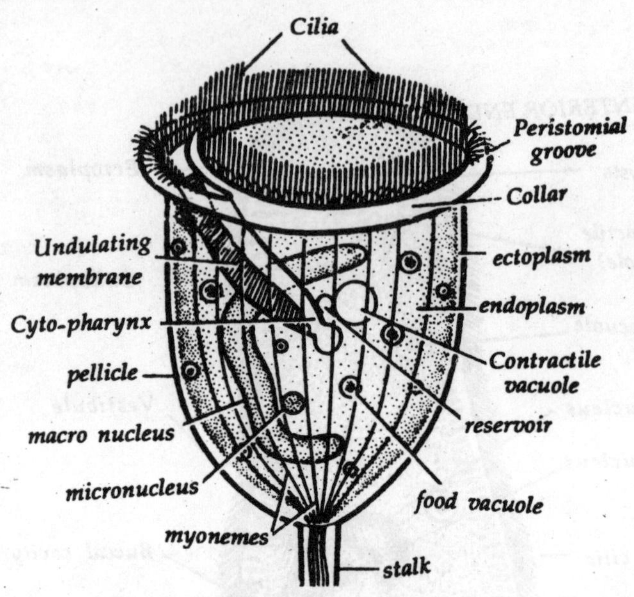

Vorticella

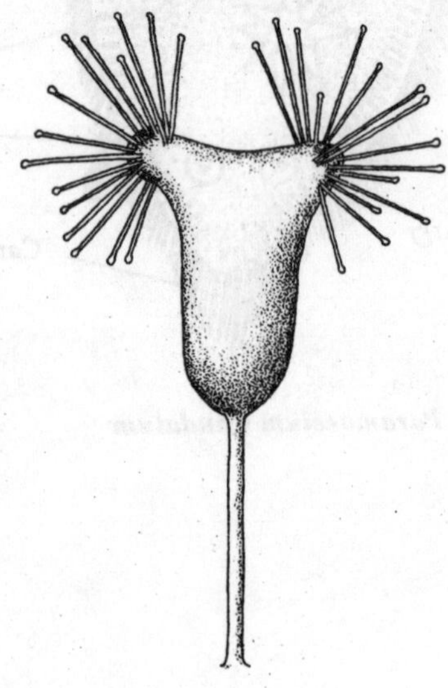

Acineta

PORIFERA

Phylum Porifera consists of the most primitive multicellular animals called sponges. They are incapable of moving from place to place and resemble plants in appearance. The phylum name Porifera (Latin. porus = pore + ferre = to bear) was coined by Robert E. Grant. Poriferans or sponges contain several tiny pores in their body walls. Hence, they are usually referred to as pore bearers. Sponges are completely aquatic and mostly marine. There are nearly 10,000 species of sponges.

Although sponges are multicellular animals, their morphology is different from other Metazoa. The cells in the body are loosely arranged and there is no indication of tissues and organs. It appears that sponges have separated from the general metazoan stalk at an early period. They are not on the direct line of evolution and did not give rise to any other metazoan group. This led to place the sponges under a separate group called Parazoa.

Leucosolenia, Sycon, Euplectella, Spongilla, etc., are some of the common examples of sponges.

GENERAL CHARACTERS

1. Sponges are the multicellular animals at the lowest level. They have cellular level of body organisation and no definite tissues or organs.

2. All are aquatic and mostly marine. A few are fresh water animals.
3. Symmetry is radial or none.
4. They are solitary or colonial and fixed (sedentary) like plants.
5. They are of different shapes: vase-like, cylindrical, tubular, globular, cushion-shaped, etc.
6. Body wall is composed of two cellular layers, hence it is called diploblastic. Outer layer of the body wall is called **pinacoderm** and the inner layer is called **choanoderm**. There is a non-cellular gelatinous **mesenchyme** in between the two layers. Mesenchyme is having free amoeboid cells and skeletal structures.
7. Sponges have many surface openings called **ostia** or inhalant pores. **Canals and chambers** are also present in the body wall through which water flows. **Flagellated cells** or **choanocytes** are present in special flagellated or radial chambers.
8. **Skeleton** is in the form of calcareous or siliceous spicules or organic spongin fibres or both.
9. All canals open into common cavity known as **spongocoel** or **paragastric cavity**. This cavity opens to the outside through and anterior aperture called **osculum.**
10. Mouth and digestive tract are absent. Digestion is intracellular.
11. There are no specialised respiratory or excretory organs or movable appendages.
12. Definite nerve cells and sensory cells are lacking.
13. All sponges are bisexual or hermaphrodites, but cross fertilization occurs. *Sex cells are formed from amoebocytes.*
14. Asexual reproduction takes place by **budding** or **gemmules**.
15. Cleavage is holoblastic. Development is indirect which includes free-swimming, ciliated larva, the amphiblastula or Paranchymula.
16. Power of regeneration is very high in sponges.

CLASSIFICATION

The phylum *Porifera* is divided into three classes, based on spicules.

Class 1. Calcarea

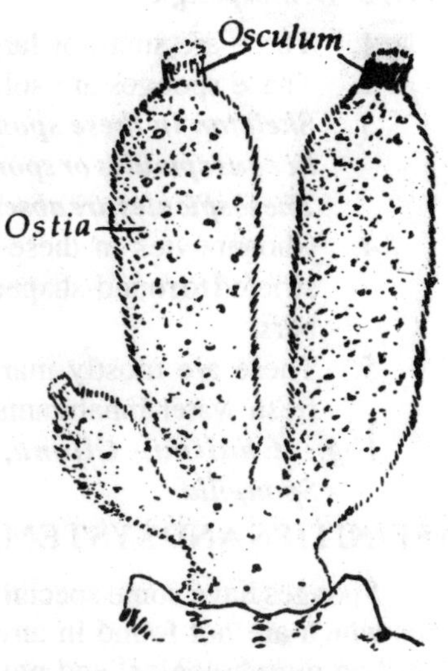

***Scypha* Colony**

1. These are small sized calcareous sponges.
2. These sponges are solitary or colonial.
3. *Skeleton in these sponges is composed of calcareous spicules.*
4. These are exclusively marine and mostly live in shallow waters.
5. Body structure of a sponge is generally simple and covered with spines.

   E.g.: ***Leucosolenia, Clathrina, Grantia, Scypha.***

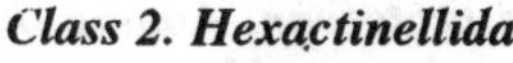
Class 2. Hexactinellida

1. These are moderate sized glass sponges.
2. Skeleton in these sponges consists of siliceous spicules.
3. These sponges are exclusively marine and many live in deep sea.
4. Choanocytes in these sponges are restricted to finger-shaped flagellated chambers.

E.g.: ***Hyalonema, Pheronema, Euplectella.***

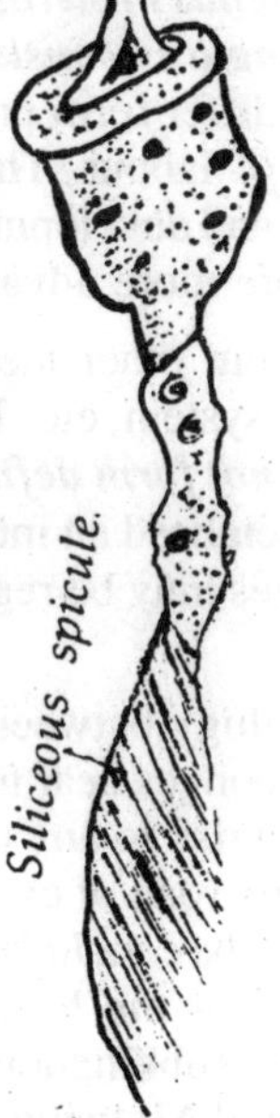

Hycalonema

Euplectella

Class 3. Demospongia

1. These are small or large sized sponges.
2. These sponges are solitary or colonial.
3. ***Skeleton in these sponges consists of siliceous spicules or spongin fibres or both. These spicules are absent in some sponges.***
4. Choanocytes in these organisms are restricted to round-shaped flagellated chambers.
5. These are mostly marine and a few are fresh water inhabitants.

E.g.: ***Chalina, Cliona, Hippospongia, Spongilla.***

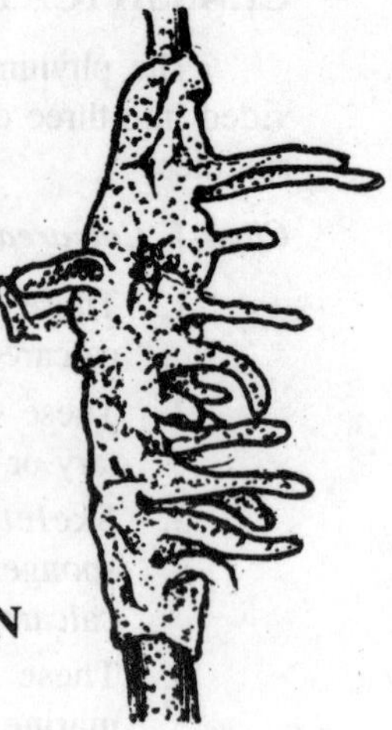

Spongilla

AFFINITIES AND SYSTEMATIC POSITION

Sponges have some special characteristic features which are not found in any other metazoans. Based on morphological and physiological considerations, it appears quite reasonable to separate them from **Metazoa** and include under a separate subkingdom, **Parazoa.**

Based on the presence of choanocytes, sponges are believed to be the animals evolved from a choanoflagellate protozoan. Among the choanoflagellates, the genus Proterospongia closely resembles a sponge. Therefore, Proterospongia is considered as the probable ancestor of sponges. But, sponges differ form protozoans in having cellular differentiation and division of labour. The presence of canal system and a characteristic skeleton, and development from fertilized egg by cleavage indicate that sponges are more advanced than protozoans.

Sponges differ from other metazoans in the absence of mouth, digestive tract, nervous system, etc. ***The cells of sponges are only loose aggregations. They do not form definite tissues.*** Moreover, the cells of sponges are far less specialised an interdependent than the cells of other Metazoa. Hence, sponges may be regarded as the multicellular animals at the lowest level.

There is no homology between the development of sponges and other metazoans. The sponges bear no real similarity to Metazoa. They certainly have not given rise to any of the other metazoan group. They are considered as on a by road of evolution and not on the main line of evolution. *Finally, it may be concluded that the sponges constitute a blind branch diverged from near the base of the evolutionary tree.* Hence, sponges are placed in the subkingdom **Parazoa,** the other two subkingdoms being **Protozoa** and **Metazoa**.

CNIDARIA

The Coelenterata or Cnidaria includes hydras, jelly fishes, sea anemones, corals, etc. The term Cnidaria was used earlier for a subphylum of the phylum coelenterata (koilos = hollow + enteron = intestine). The term Coelenterata was coined by Leuckart. However, Barnes, in 1980, used the term Cnidaria synonymous to the phylum Coelenterata.

Cnidarians show radial symmetry. They exhibit only tissue level of organisation so definite organs are lacking. The cells are arranged in two layers, hence they are called diploblastic animals. The cells perform different functions, exhibiting division of labour.

Cnidarians are true multicellular animals belonging to a separate sub-kingdom, the Metazoa. All animals are aquatic and none of them are parasitic in nature. Most of them are marine and very few like Hydra are fresh water animals.

There are about 11,000 species of cnidarians. They are either solitary or colonial and exist in two different forms namely polyp and medusa. Polyp is cylindrical in shape, usually found attached to some solid substratum whereas medusa is an umbrella-shaped free swimming form. Polymorphism is common in many coelenterates. Hydra, Obelia, Aurelia, Metridium are some of the common examples of the phylum.

GENERAL CHARACTERS

1. All Cnidarians are aquatic. Some live in fresh water and most of them are marine.

2. *They show tissue grade of body organisation, i.e., the cells are specialized for different functions.*
3. These are solitary or colonial, sedentary or free-swimming.
4. There is no head. Segmentation is lacking.
5. *These are radially or biradially symmetrical animals with a longitudinal oral-aboral axis.*
6. Two types of individuals namely **polyp** (sessile or free) and **medusa** (free swimming) are present.
7. Body wall is diploblastic consisting of two cellular layers, outer **epidermis** and inner **gastrodermis**. In between these two layers, there is non-cellular gelatinous mesoglea. In some, the mesoglea contains cells and connective tissue.
8. Special stinging cells called **cnidoblasts** are present in one or both cellular layers. *The phylum Cnidaria derives its name due to the presence of cnidoblasts.* They are especially abundant on tentacles.
9. The inner cavity is known as **gastrovascular cavity** or **enteron** or **coelenteron.** In some animals it is branched and decided by septa or mesenteries. Coelenteron opens outside through an opening called mouth. Anus is absent.
10. Usually the mouth is surrounded by long thread-like tentacles. They assist in food capture and locomotion.
11. *Digestion is extracellular as well as intracellular.*
12. Circulatory, respiratory and excretory systems are absent.
13. Nervous system consists of nerve cells. The nerve fibres of these cells are inter-connected to form a diffuse **nerve plexus** or **nerve net**. There is no brain, hence nervous system is of primitive type. Sense organs are in the form of **ocelli** or **statocysts**.
14. Asexual reproduction occurs by budding or fission. Sexual reproduction takes place by gametes. Both unisexual and bisexual animals are present.
15. Alternation of generation is seen in the life history of some animals.
16. Development usually included a free swimming ciliated larva called planula.

CLASSIFICATION

The phylum cnidaria is divided into three classes, 1. Hydrozoa , 2. Scyphozoa and 3. Anthozoa, based on zooids and nature of mesoglea.

Class 1. HYDROZOA

1. The members are solitary or colonial, sessile or free living animals; both unisexual as well as bisexual in nature.
2. They live both in fresh and marine waters, but most of them live in sea water.
3. Both polyp and medusa stages are present, but polyp stage is predominant.
4. Gastrovascular cavity is simple without mesenteries.
5. Medusae are with true velum.
6. Gonads and cnidoblasts are ectodermal in origin.

Examples : ***Hydra, Obelia, Physalia, Velella, Porpita.***

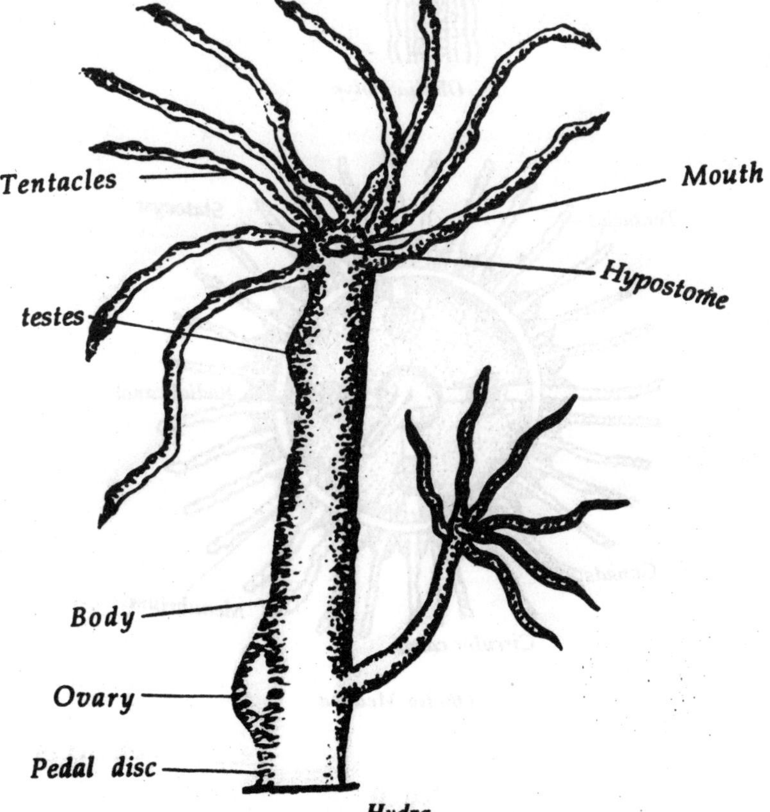

Hydra

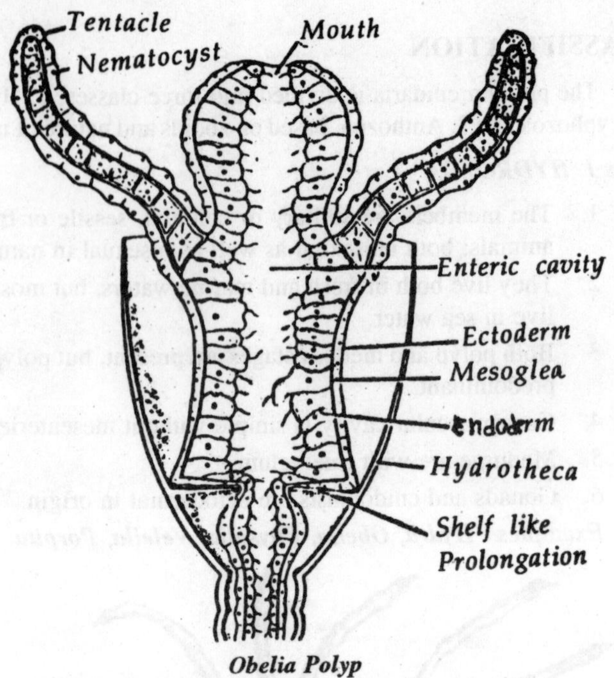

Obelia Polyp

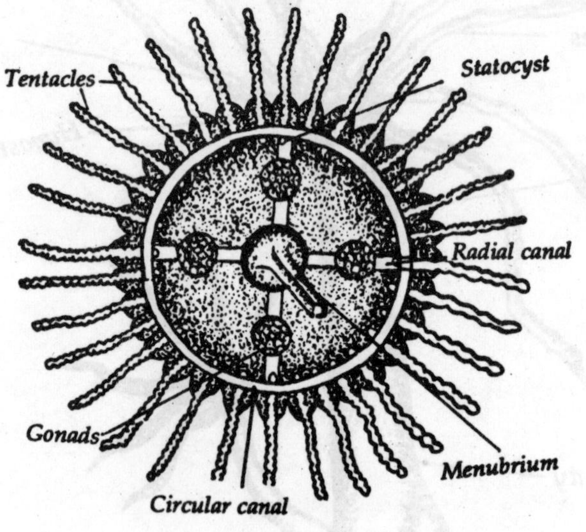

Obelia Medusa

Class 3. ANTHOZOA or ACTINOZOA

1. The members of this class are exclusively marine. Both unisexual and bisexual animals are present.
2. These are solitary or colonial. Sea anemones and coral animals are included in this class.
3. All are polyps, no medusae.
4. Mouth leads into pharynx or stomodaeum.
5. Gastrovascular cavity is divided by septa or mesenteries.
6. Gonads and cnidoblasts are derived from endoderm and they are seen on mesenteries.

Examples: ***Adamsia, Metridium, Alcyonium, Gorgonia Meandrina, Fungia.***

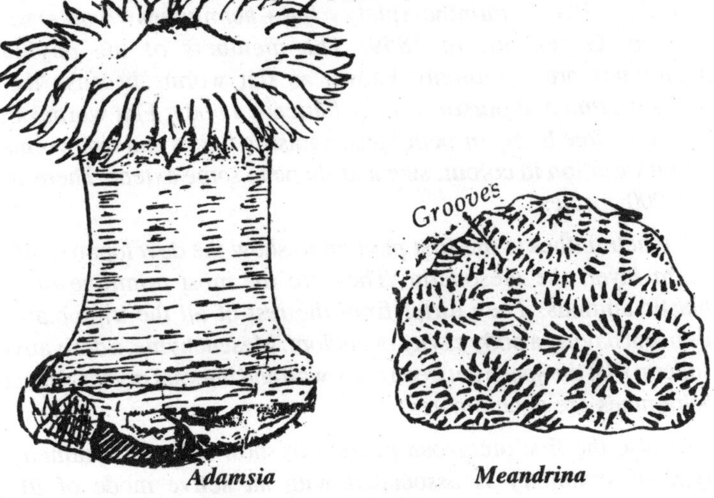

Adamsia ***Meandrina***

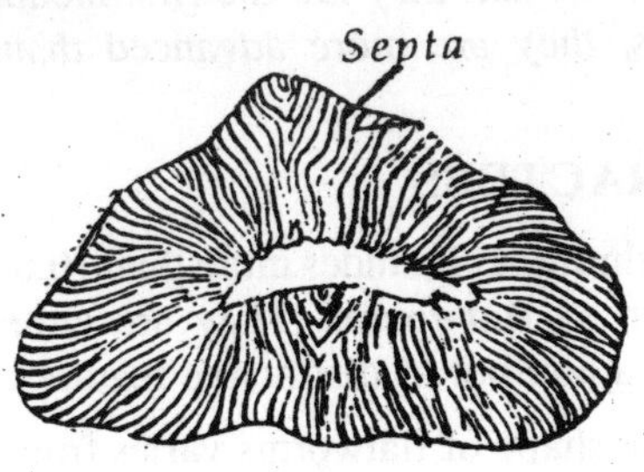

Fungia

PLATYHELMINTHES

The name Platyhelminthes (platy = flat + helminthes = works) was proposed by Gegenbaur in 1859. The members of the phylum Platyhelminthes are commonly known as flat worms because they possess a soft, thin and dorsoventrally flattended body. Flat worms are represented by free living turbellarians, parasitic flukes and tapeworms. They show variation in colour, size and shape to some extent. There are about 15,000 species.

Platyhelminthes is the first phylum to show the development of a third germ layer, the mesoderm. They are the most primitive of all triploblastic animals. They differ from the rest of all the triploblastic animals in not having a body cavity or coelom. Mesenchyme, a derivative of mesoderm fills the space between body wall and internal organs. Hence Platyhelminthes are acoelomates.

It is also the first metazoan phylum to show bilateral symmetry. This type of symmetry is associated with an active mode of life. Flatworms are unique, in that they are the first metazoans to develop organ systems. Thus, they are more advanced than the Radiata or Diploblastica.

GENERAL CHARACTERS

Phylum Platyhelminthes includes more than 10,000 species which are commonly known as **flatworms**. They are either free living or parasitic. They show a number of distinctive features.

1. **Shape :** The shape of flatworms varies from leaf-like bodies

to ribbon-like bodies.

2. **Size :** Flatworms also show a lot of variation in size. Some free living flatworms are less then 50 mm. in length, whereas, some tapeworms are more than 20 meters in length.
3. **Colour :** Free living flatworms show brilliant colours. But, most of the parasitic flatworms are either white or colourless. Some parasites may get their colour from the ingested food.
4. **Triploblastic condition :** In addition to the two germ layers, ectoderm and endoderm, the third germ layer, mesoderm is also formed in flatworms. Hence, they are said to be the first triploblastic animals.
5. **Bilateral symmetry :** Flatworms are the most primitive group of bilaterally symmetrical forms. Body parts are arranged laterally on either side of the main axis. The body had distinct anterior and posterior ends, dorsal and ventral surfaces.
6. **Exo- and endo-skeleton :** These are completely absent. Therefore, the body is usually soft.
7. The hard parts are in the form of spines, hooks, cuticle, etc.
8. A single layered ciliated epidermis is present in turbellarians. In flukes and cestodes, the epidermis is absent and the body is covered by a tegument.
9. **Orangs of attachment :** The parasitic forms have developed suckers, spines, hooks, etc., as organs of attachment to the tissues of the host.
10. **Acoelomate grade :** All flatworms are without a body cavity or coelom. Hence, they are termed as acoelomates. This is the most distinguished feature of flatworms.
11. **Mesenchyme or Parenchyma :** It is a loose, spongy tissue filling the space between body wall and internal organs. It is formed from mesoderm.
12. **Muscles :** Both circular and longitudinal smooth muscles are well developed in the body of flatworms.
13. **Organ system level of development:** Flatworms represent the first metazoan group which developed organ systems.
14. **Digestive system :** It consists of mouth, pharynx and a branched intestine which ends blindly without anus. The digestive system is completely absent in tapeworms.
15. **Respiratory and circulatory systems** are absent. Respiration

in endoparasitic flatworms is **anaerobic.**

16. **Nervous system :** It is of primitive type. It consists of a pair of cerebral ganglia or brain and 3 to 5 pairs of longitudinal nerve cords. These nerve cords are connected by transverse commissures. *Thus a ladder-like nervous system is formed in majority of flat worms.*
17. **Excretory system :** It consists of longitudinal excretory canals bearing number of flame cells. The flame cells are the organs of excretion and osmoregu-lation in flatworms.
18. Sense organs are usually absent or reduced in parasitic flat-worms. But, they are found in tubellarians in simple form.
19. Flatworms are hermaphrodites with some exceptions. Copu-latory organ is present. Germarium (ovary) and yolk producing vitellarium are separate.
20. Fertilization in internal.
21. Development is indirect with one or more larval stages.
22. Life cycle is complicated, often involving two or more hosts.

CLASSIFICATION

Phylum Platyhelminthes is divided into 3 classes, viz., Turbellaria, Trematoda, and Cestoda.

CLASS:TURBELLARIA

(turbella = a little string)

1. This class includes about 1500 species which are commonly known as **planarians.**
2. The members of this class are mostly aquatic, living in fresh or salt water. A few forms live in moist soil. Land planarian (*Bipalium*) is the largest of all turbellarians.
3. There is a distinct heat at the anterior end.
4. The body is covered with a ciliated epidermis which contains gland cells and rhabdites. The rhabdites are hyaline rod-like bodies which disintegrate into slime.
5. The pharynx comes out in the forms of proboscis during ingestion.
6. Most of the turbellarians are carnivorous forms.
7. Sense organs are present in the form of ciliated pits, grooves, simple eyes, etc.

8. Cleavage is spiral and determinate type.
9. Planarians have remarkable power of regeneration.

Examples : ***Dugesia, Bipalium***

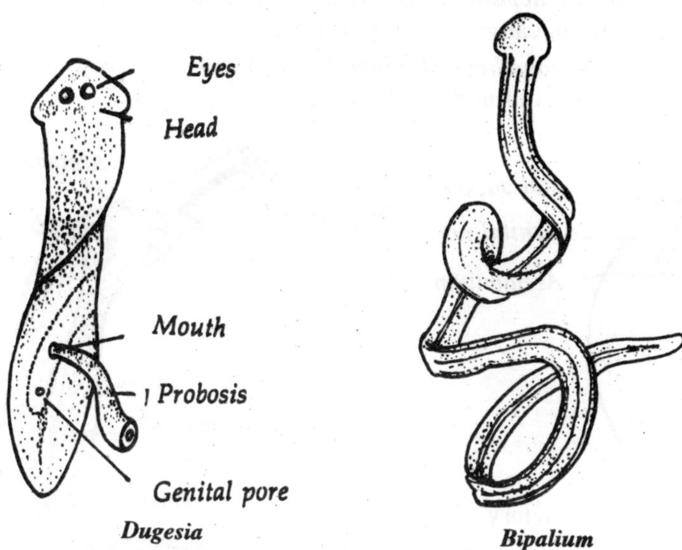

Dugesia ***Bipalium***

CLASS : TREMATODA
(Gr. trema = apertures)

1. The class Trematoda has about 6000 species which are commonly known as **flukes.**
2. Trematodes are either ecto-or endo parasites.
3. The body is oval or elongated and unsegmented. The thin fl... body helps the flukes to adjust themselves in the minute tiss... spaces of the host.
4. The body of trematodes generally bears two suckers, name... oral and ventral suckers.
5. Epidermis is absent in the body of flukes.
6. The body is covered with non-ciliated but spined tegume...
7. Life cycle of endoparasitic flukes involves two or more hos...
8. They also cause diseases in man, domesticated animals a... fish.

Examples

1. ***Gyrodactylus*—**an ectoparasite on fish
2. ***Fasciola hepatica*** (Liver fluke)—a pathogenic endoparasite of sheep, causes liver rot in sheep.
3. ***Schistosoma haematobium*** (Blood fluke)—lives in the blood capillaries of urinary bladder of man.

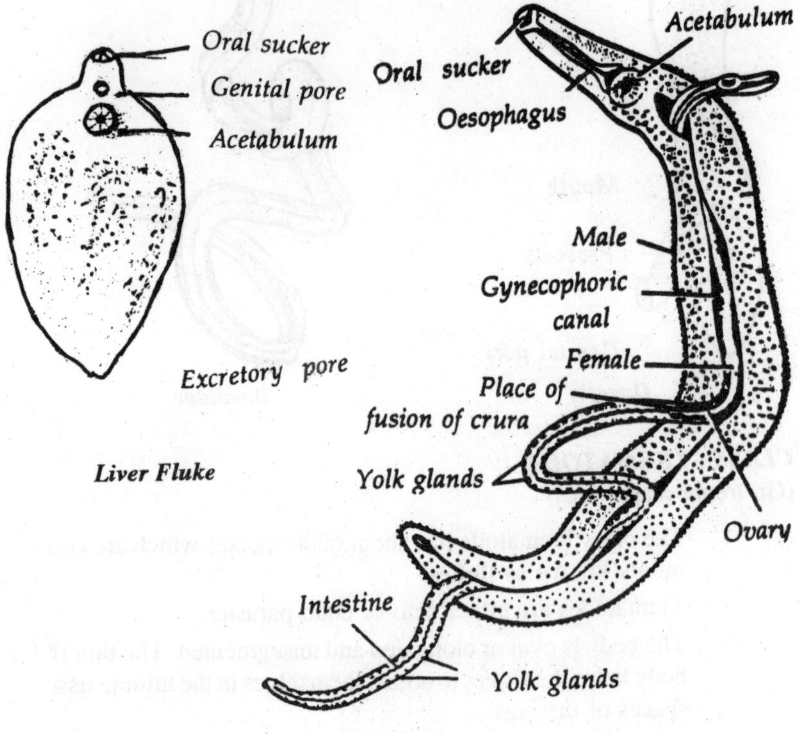

Liver Fluke

Schistosoma haematobium

CLASS : CESTODA

(cestus = tape + eidos = form)

1. The class Cestoda consists of about 3,400 species which are commonly known as **tapeworms.**
2. Cestoda is the most highly specialised class of phylum Platyhelminthes.
3. These are segmented flatworms.
4. Cestodes are exclusively endoparasites and show distinct

parasitic adaptations.

5. The flat ribbon-shaped body is divisible into scolex neck and strobila.
6. The strobila is generally composed of 3 types of proglottids.
7. The body cestodes is also covered with a non-ciliated tegument but it bears microvilli.
8. The digestive system is totally absent.
9. Life cycle of cestodes involves two or more hosts.

The class cestoda is divided into 2 subclasses viz., **Cestodaria** and **Eucestoda.**

1. Subclass CESTODARIA :

The cestodarians differ from the rest of the cestodes in the absence of scolex and strobila. They resemble trematodes very closely.

*Example **Gyrocotyle*** - parasite on sharks

2. Subclass EUCESTODA :

The eucestodes constitute the majority of cestodes. Body is divisible into scolex neck and strobila.

*Example : **Taenia solium, Taenia saginata, Dibothriocephalus latus***

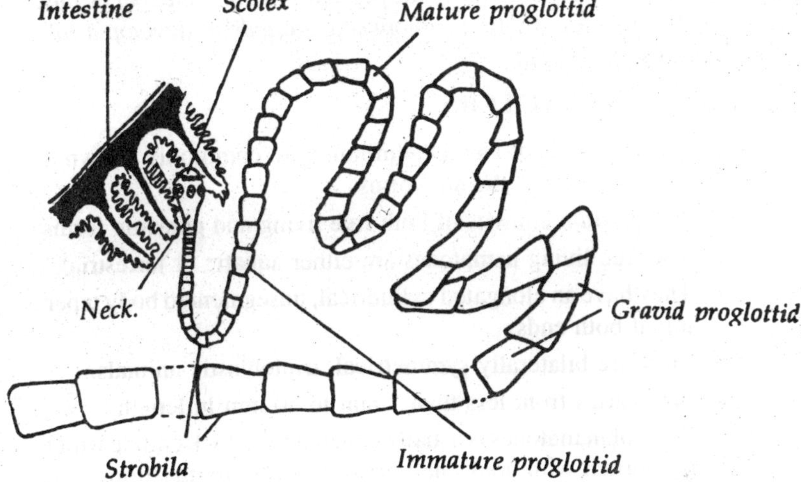

Taenia solium

NEMATHELMINTHES

The term Nemathelminthes (Gr. nematos = thread + helminthes = worms) was proposed by Gegenbaur in 1851. It includes a diverse group of pseudocoelomate animals. Barness 1980 included phylum Nemathelminthes or Nematoda along with Rotifera and Gastrotricha in Aschelminthes (Gr. askos = hollow; helminthes = worm). There are about 10,000 species in Nematoda.

The members of this phylum are commonly known as hollow worms (round worms or thread worms). They are triploblastic, bilaterally symmetrical and unsegmented worms. They are distinguished from the rest of the triploblastic animals by the presence of a pseudocoel or false body cavity. They are the first triploblastic animals to develop a tube within the tube body plan.

GENERAL CHARACTERS

1. The members of Nemathelminthes are commonly known as round worms or thread worms.
2. The phylum consists of both free living and parasitic forms.
3. The free living nematodes are either aquatic or terrestrial.
4. They have an elongated cylindrical, unsegmented body tapering at both ends.
5. They are bilaterally symmetrical, triploblastic animals.
6. Size varies from less than 1 mm to 50 mm in length.
7. Body of nematodes is usually covered by a thick cuticle which is adorned with hooks and bristles in free living forms.

8. Epidermis of nematodes is a syncytium. It forms 4 longitudinal lines or chords.
9. There is only longitudinal muscle layer. Circular muscle layer is absent.
10. Body cavity of nematodes is a pseudocoel which is a remnant of blastocoel. It is one of the distinguished features of nematodes.
11. Digestive system consists of a straight alimentary canal with mouth, pharynx, intestine and anus. But, there are no digestive glands.
12. Excretory system consists of intracellular canals. Flame cells are absent.
13. *Nematodes show limited and constant number of cells in various tissues. This condition is known as entity.*
14. Nervous system consists of a cerebral nerve ring and longitudinal nerve cords.
15. Sense organs are poorly developed in parasitic nematodes. They are in the form of papillae near the 2 ends of body.
16. Circulatory and respiratory systems are absent.
17. Sexes are separate (Dioecious /unisexual). There is distinct sexual dimorphism. Males are smaller than females and possess curved tails with penial spicules.
18. Gonads are in the form of coiled tubes.
19. Fertilization is internal.
20. Cleavage is often spiral and determined type.
21. Development is indirect.
22. Majority of nematodes are oviparous. Eye worms, filarial worm, trichina worm, etc. are viviparous.
23. They are either monogenetic (Ascaris, pin worm, Hook worm) or digenetic (Filarial worm, Eye worm).
24. There are 4 larval stages with 4 moultings. The fifth stage is the adult.
25. Nematoda is the second largest class in the animal kingdom.

CLASSIFICATION

Nematoda is divided into 2 subclasses on the basis of presence or absence of phasmids.

Sub class : APHASMIDA

Phasmids are absent. There are no excretory canals. Most of them are free living forms.

Example

Enoplus, Trichenella, Trichiuris

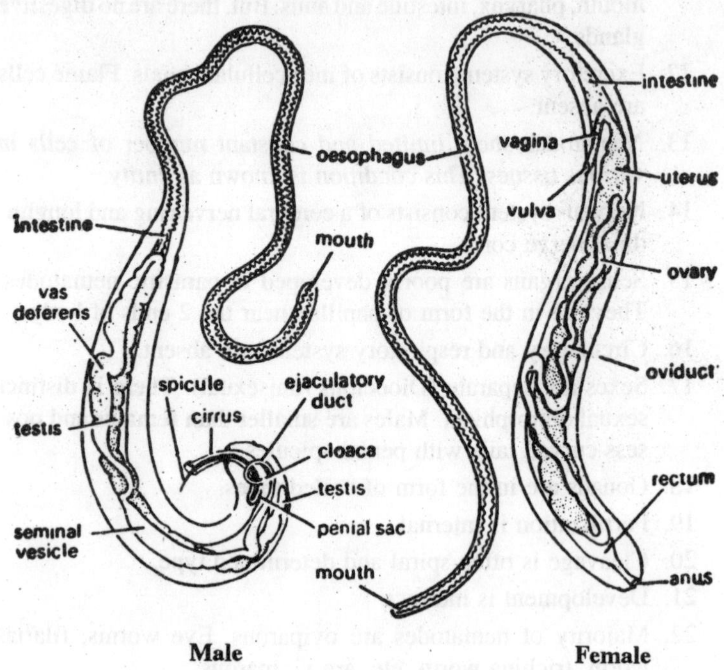

Trichuris trichiura

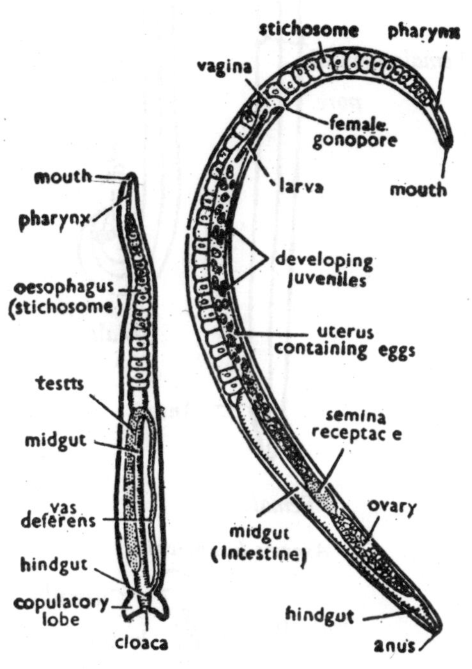

Male **Female**

Trichinellasspiralis

Sub class : PHASMIDA

Phasmids and excretory canals are present. Most of them are parasites.

Example

Ascaris lumbricoides, Enterobius or Oxyuris vermicularis, Ancylostoma duodenale, Wuchereria bancofti.

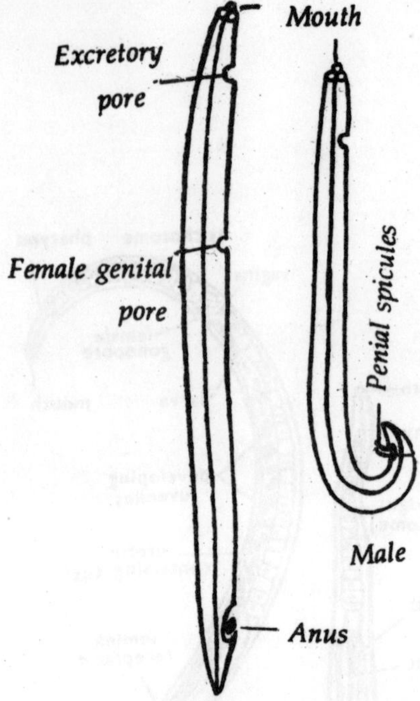

Ascaris lumbricoides

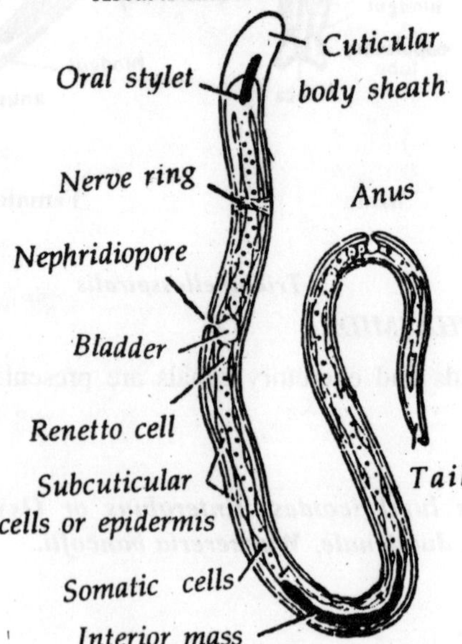

Wuchereria

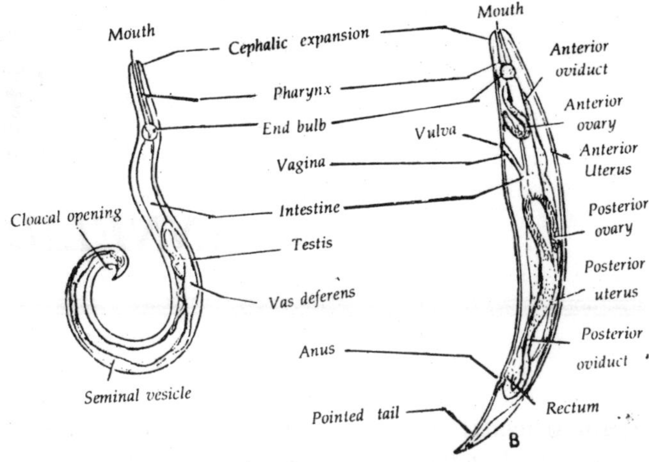

Enterobius vermicularis

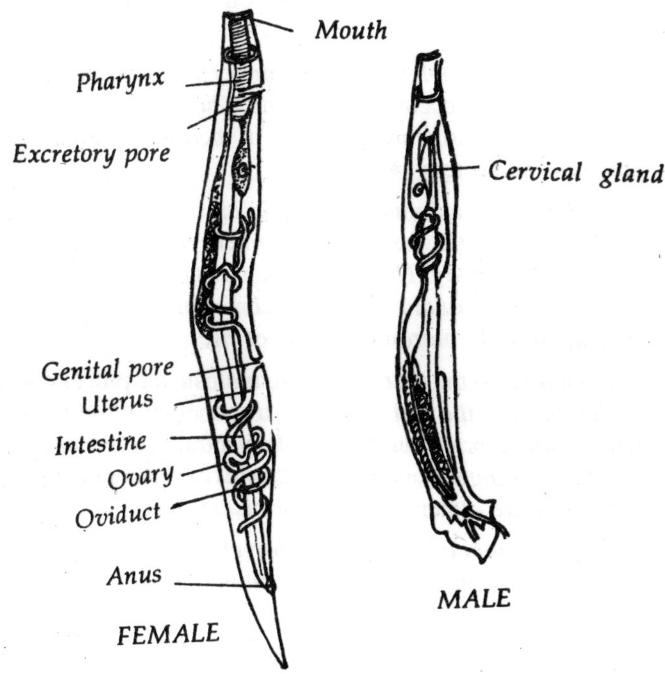

Ancylostoma

ANNELIDA

The term Annelida (L : annulus = ring; Gieidos = form) was coined by Lamarck in 1809. Annelids are commonly known as higher segmented worms. They are worldwide in distribution. Annelids are highly advanced and the most successful of the various worms. Formerly, all worms are included in the group vermes. Later, Cuvier separated the segmented worms from the rest.

Annelids are triploblastic, bilaterally symmetrical, metamerically segmented and true coelomate animals. They show a lot of diversity in habit, habitat and structure. Majority of them are free living, some are commensals and a few are parasites.

The body of an annelid is made up of a linear row of ring-like segments called metameres. Each metamere possesses representatives of all systems of the body. This kind of segmentation is known as metamerism which is the most distinguishing character of annelids. It is evolved as an adaptation to burrowing mode of life.

Annelids are the first animals to indicate the process of cephalization. They are also the first to develop true coelom, closed blood vascular system, distinct brain and a double ventral nerve cord. They have segmentally arranged coiled, citiated tubes called nephridia as excretory organs. They are either unisexual or bisexual.

Phylum annelida includes polychaetes, earthworms, leeches archiannelids and myzostomids amounting nearly 7,000 species.

GENERAL CHARACTERS

1. Annelids are mostly marine; some are fresh water, some are terrestrial and a very few are ecto parasites.
2. They are either burrowing or crawling or tube dwelling forms.
3. They possess a vermiform elongated body.
4. *The symmetry is bilateral in annelids* as the body parts are arranged on either side of the ora-anal or median axis of the body. The right and left halves are similar like mirror images.
5. They are triploblastic animals because there are three primary germ layers-ectoderm, endoderm and mesoderm- in the embryonic stage. All organs arise from these germ layers.
6. Metamerism is the chief distinguishing characters of annelids. Body is divided into metameres externally by circular grooves and internally by muscular septa. In some forms, the anterior segments form a distinct head (cephalisation).
7. There is a thin, non-chitinous irridescent cuticle as the outer most covering of the body.
8. Body wall has well developed longitudinal and circular muscles to facilitate peristaltic contractions. It is said to be **dermo-muscular.**
9. Segmentally repeated groups of chitinous bristles called setae are found except leeches and archiannelids. Setae are the organs of locomotion in Annelida.
10. *Body cavity of annelids is a true coelom as it is lined with coelomic epithelia.* It is filled with coelomic fluid which functions as a hydraulic skeleton. However, coelom in leeches is reduced due to the development of botryoidal tissue.
11. There is a straight alimentary canal beginning with mouth at the anterior end and terminating in an anus at the posterior end. *Digestive glands are developed first in Annelida. Digestion is intercellular or extracellular.*
12. Usually, there are no organs of respiration. It is performed through moist skin (cutaneous). Gills are present rarely in some polychaetes (Arenicola).
13. *Annelids are the first animals to have a closed type of blood vascular system.*
14. The red color of blood in annelids is due to the presence of the dissolved haemoglobin in plasma. There are no erythrocytes.

15. Excretory organs are segmentally repeated coiled ciliated ectodermal tubes generally known as **nephridia.**
16. Excretory product of aquatic forms is ammonia and that of semi-terrestrial forms is urea.
17. The nervous system is well developed. It comprises of a distinct brain and a double ventral nerve cord.
18. *Gonads are mesodermal in origin. They are proliferated from coelomic epithelia.*
19. They are either unisexual (polychaetes) or bisexual (oligochaetes and leeches).
20. Fertilization is external except in leeches.
21. Cleavage is spiral and determinate type.
22. Development is indirect in polychaeta and Archiannelida, but direct in earthworms and leeches.

CLASSIFICATION

Phylum Annelida includes about 10,000 species which are worldwide in distribution. It is divided into 5 classes viz.,

1. Polychaeta
2. oligochaeta
3. Hirudinea
4. Archiannelida
5. Myzostomida.

Polychaeta and oligochaeta are together referred to as **Chaetopoda** because they posses setae.

Class–Polychaeta

(Gr. poly = many; chaete = setae)

1. It is the largest class of annelida with over 7,000 species.
2. They are predominantly marine forms with a distinct head bearing simple eyes, palps and tentacles.
3. There are numerous setae born on lateral feet called parapodia which are biramous appendages. *The parapodia are the organs of locomotion in Polychaeta.* They are not found in the rest of the annelids.
4. They are mostly burrowers, some are crawlers or tube dwellers or planktonic forms.
5. There is no clitellum.
6. Polychaetes are unisexual or dioecious.
7. Gonads are numerous and develop only during the breeding

season. There are no gonoducts.

8. Sex cells are liberated by the rupture of the body wall of the posterior half.
9. Development is indirect as there is pear-shaped **Trochophore larva** in the life history.
10. They show adaptive radiation.

Class polychaeta is divided into two sub-classes *viz.*, **Errantia** and **Sedentaria.**

Sub-class Errantia (errars = wandering)

These are burrowing or swimming or crawling or pelagic polychaetes with a distinct head and well developed parapodia.

Examples : ***Neries, Neanthes, Polynoe, Aphrodite, Glycera, Eunice***

Sub-class Sedentaria

1. These are tube swelling polychaetes. They feed on detritus. 2. Head is small with indistinct prostomium. 3. There is a nonprotrusibile pharynx with jaws.

Examples : ***Arenicola, Chaetopterus, Sabella.***

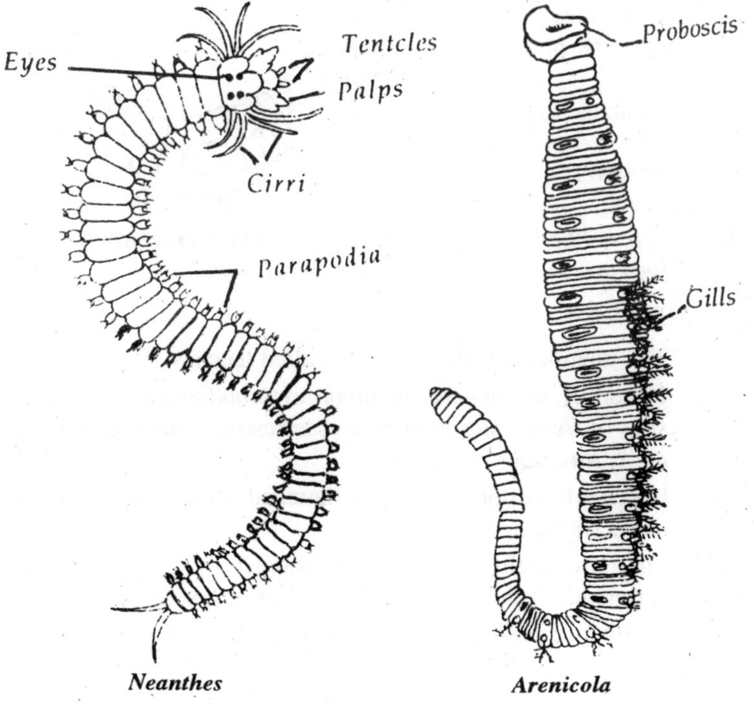

Neanthes ***Arenicola***

Class–Oligochaeta

(oligos = less, chaete = setae)

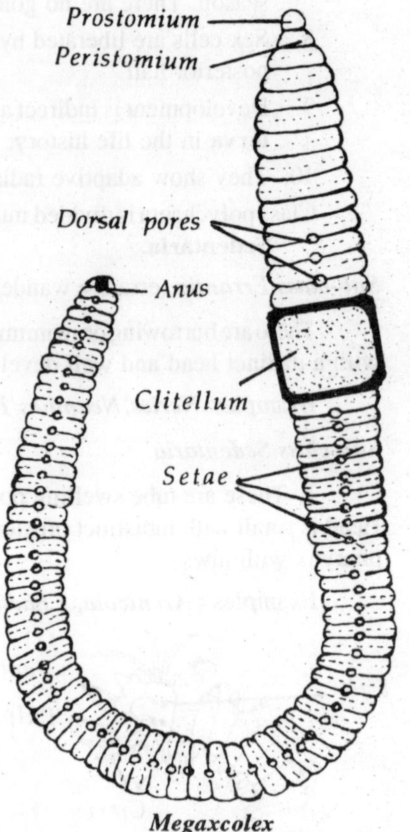

Megaxcolex

1. It is the second largest class of Annelida.
2. They are mostly semi-terrestrial and some are fresh water forms.
3. Head and parapodia are absent.
4. Small number of setae are embedded in setigerous sacs of the body wall.
5. Clitellum is formed in mature condition.
6. All are hermaphrodites or bisexuals.
7. Permanent gonads with gonoducts are present.
8. Development is direct as there is no larval stage.
9. They have the power of regeneration.
10. Some oligochaetes reproduce asexually by fission (***Chaetogaster***).

Examples:

Megascolex, Pheretima, Lumbricus, Drawida, Octochaetus, Tubifex (blood worm), Allolabophora and Aelosoma.

Class–Hirudinea

1. The members of this class are called leeches.
2. *Generally, lecches are sanguivorous (blood sucking) ectoparasites of fresh water or marine or terrestrial vertebrates. Some are carnivorous.*
3. Leeches have dorso-ventrally flattened, elongated and highly elastic bodies.
4. They are distinguished by the absence of head, setae and parapodia. Anterior part bears 5 pairs of dorsal simple eyes.
5. Number of body segments is definite, usually 33. These are

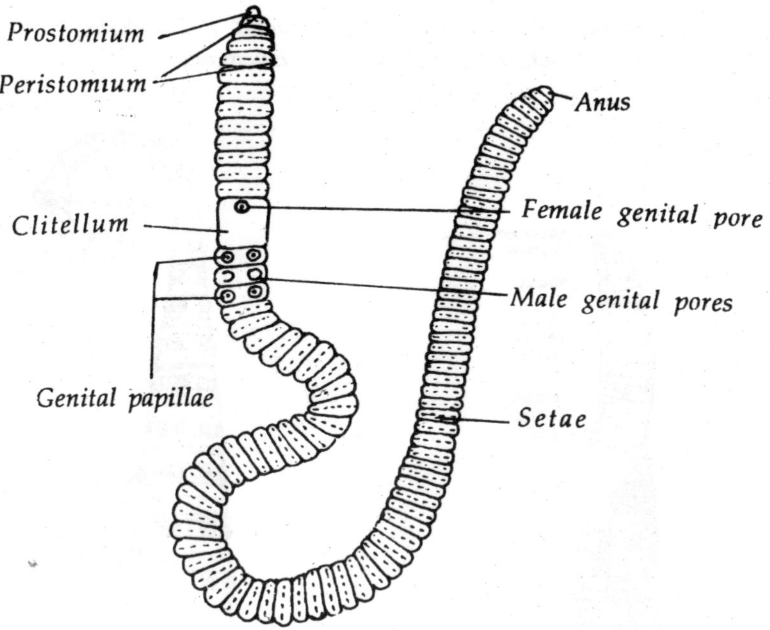

Pheretima

subdivided into annuli. In Indian cattle leech, the number of annuli is 109.

6. They possess two ventrally directed anterior and posterior suckers at anterior and posterior ends of the body respectively. The anterior suckers serve as organs or locomotion and attachment.
7. The coelom is very much reduced by the development of botryoidal tissue or mesenchyme.
8. *Circulatory system is of open type and is called haemocoelic system. It contains red haemocoel.*
9. They are bisexual or monoecious and develop clitellum only during breeding season.
10. Leeches differ from the rest of the annelids in having a copulatory organ.
11. Fertilization is internal and development takes place inside the cocoon.

12. Development is direct as there is no larval stage.

Examples : ***Hirudinaria, Acanthobdella, Pontobdella, Ozobranchus, Hirudo, Haemadipsa.***

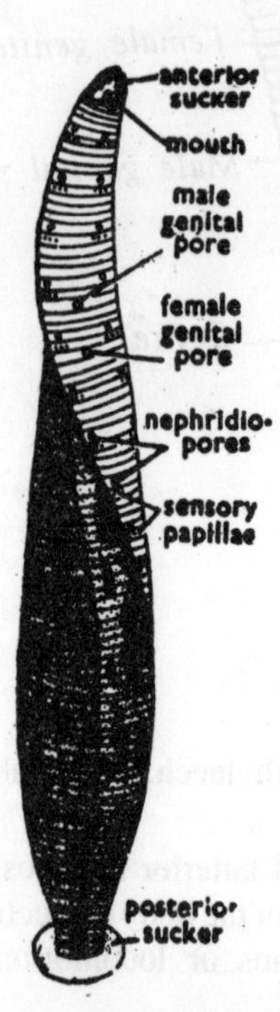

Hirudo medicinalis

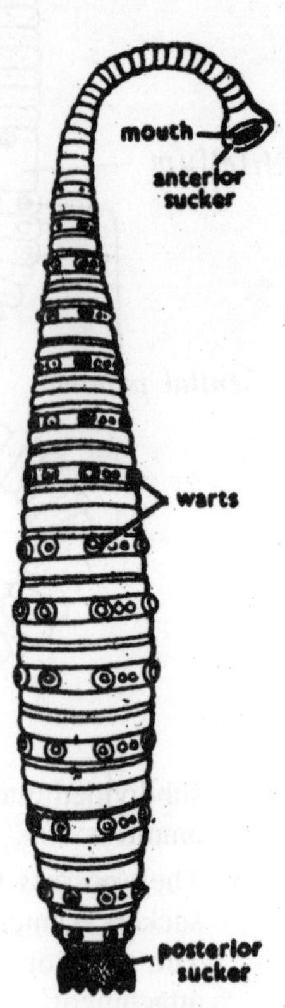

Pontobdella

Class–Archiannelida

(archi=first)

1. It is a heterogenous group of simple primitive annelids.
2. They are mostly marine.
3. There is a distinct head with tentacles and palps.
4. The body bears a ciliated epidermis.
5. Internal organs are very much reduced.
6. They are generally unisexual with numerous gonads. Gonoducts are absent.

7. The larva of archiannelids is called Loven's larva (modified Trochophore).

Example : ***Protodrillus, Polygordius.***

Class–Myzostomida

1. *It is a group of minute doubtful annelids which are ectoparasitic on crinoids.*
2. They are oval, flattened and unsegmented worms with 5 pairs of parapodia.
3. Distinct coelom and circulatory system are absent.
4. The body bears 4 pairs of suckers and several tentacles.
5. They are monoecious. Penis is also present.

Example : ***Myzostoma cubunum.***

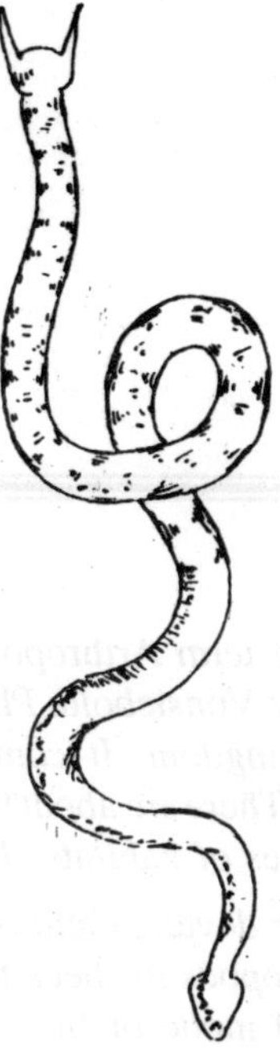

Polygordius

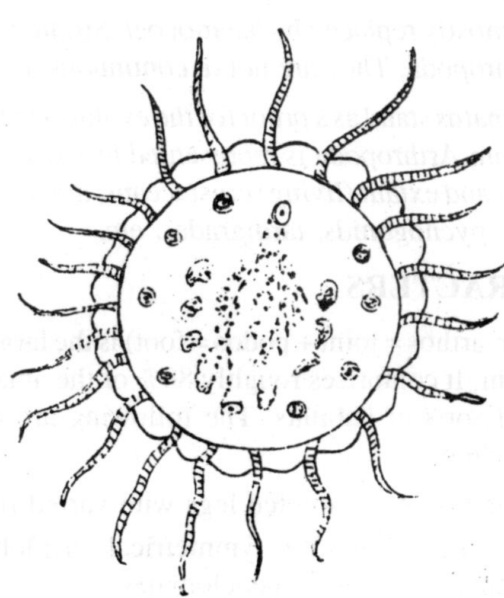

Myzostoma

ARTHROPODA

The term Arthropoda (Gr: arthros = jointed; podos = foot) was coined by Vonsiebold. Phylum Arthropoda is the largest phylum in the animal kingdom. It comprises of approximately 80% of all known animals. There are about 9,00,000 species in arthropoda. Arthropods live in all types of habitats. They show adaptive radiation.

The development of a chitinous exoskeleton enables majority of the Arthropods to check the loss of water in adapting themselves to the terrestrial mode of life. This exoskeleton is cast off several times to facilitate growth and to remove certain waste products.

The coelom is mostly replaced by haemocoel. Striated muscles are appeared first in Arthropoda. They are not in continuous muscle layers.

Forms like Peripatus stand as a proof for the evolution of arthropods from Annelids. Phylum Arthropoda is represented by extinct forms like trilobites, eurypterids and extant (living) crustaceans, arachnids, insects, miriapods, peripatus, pycnogonids, tardigrades, etc.

GENERAL CHARACTERS

Arthropoda (Gr. arthos = joint + podos = foot) is the largest phylum of the animal kingdom. It comprises roughly 80% of the animal species occupying almost all sorts of habitats. The following are the general characters of this phylum.

1. These animals possess jointed legs with varied functions.
2. Arthropods are bilaterally symmetrical, triploblastic and metamerically segmented eucoelomates.

3. Body is divisible into head, thorax and abdomen. In majority of forms, the head and thorax fuses to form cephalothorax.
4. The metamerically segmented body is covered dorsally by tergal and ventrally by sternal plates. These plates are jointed by pleura. The region between the segments of the body has a flexible cuticle for easy movements.
5. *Arthropods shed off their exoskeleton at intervals. This phenomenon is called ecdysis.*
6. Coelom in this group is a haemocoel. Blood is colourless.
7. *Mouth parts in arthropoda are the modified cephalic segments.* Mouth parts in different forms differ to suit to their food habits.
8. Digestive system is will developed with digestive glands like salivary glands.
9. *Circulatory system is open type with a dorsal heart, arteries and blood sinuses.* The heart is tubular, and chambered. It is separated from the body cavity by a pericardial membrane.
10. The excretory organs in arthropods are constituted by green glands and malpigian tubules.
11. Striated muscles are seen in Arthropoda.
12. The nervous system in this group is formed by a dorsal nerve ring and a double ventral nerve cord.
13. The sensory organs in this phylum is constituted by simple eyes (ocelli), compound eyes, chemoreceptors and tactile receptors (antennae).
14. Sexes are separate and sexual dimorphism is often exhibited by these forms.
15. Development is direct or indirect includes metamorphosis. In the development, many larval stages are also seen. *Parthenogenesis is exhibited by some organisms like insects and crustaceans.*

CLASSIFICATION

Phylum arthropoda shows three important subphyla, viz., **1. *Subphylum Onychophora*, 2. *Subphylum Chelicerata*** and **3. *Subphylum Maudibulata.***

Subphylum ONCYCHOPHORA

1. Peripatus is the only genus that is included in this subphylum. It exhibits the characters of both annelids and arthropods.

2. It has a soft skin, simple eyes, one pair of clawed appendages present in each segment, nephridia, straight alimentary canal and absence of true head. These characters are present in Onychophora. ***Peripatus*** draw close affinity to the annelids.
3. Contrary to this, ***Peripatus*** draws equal affinity to the arthropods in the features such as presence of antennae, haemecoel, dorsal tubular heart, trachael respiration and the formation of mandibles from the cephalic appendages.
4. Peripatus exhibits discontinuous distribution.

Thus Onychophora with its single representative, Peripatus, forms a link between Phylum Anelida and Arthropoda. The subphylum is also treated as a phylum by some authors.

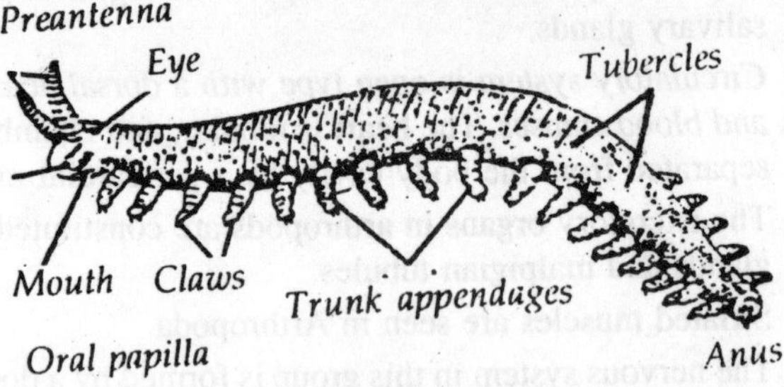

Peripatus

Sub phylum CHELICERATA

1. The body of the members of the subphylum Chelicerata (Gr. chele = claw, keors = horn, ata = group) is divisible into a cephalothorax (prosoma) and an abdomen (opisthosima). The prosoma has six segments and opisthosoma has thirteen segments.
2. The first pre-oral pair of appendages are ***chelicerae*** with claws meant for feeding. The first post-oral appendages are ***pedipalps.***
3. This group is devoid of antennae.
4. They are terrestrial or marine arthropods.

This subphylum is divided into two classes. They are ***Merostomata*** and ***Arachnida.***

Class Merostomata

1. The members of this class are totally aquatic living in marine bottoms.
2. The members show lateral compound eyes.
3. The prosoma bears 6 pairs of appendages.
4. The opisthosoma is divided into mesosoma and metasoma. The first pair of appendages in mesosoma are modified as genital operculum.
5. Metasoma is ending into a telson (Horse shoe or King crab).

Examples

Limulus, Eurypterus.

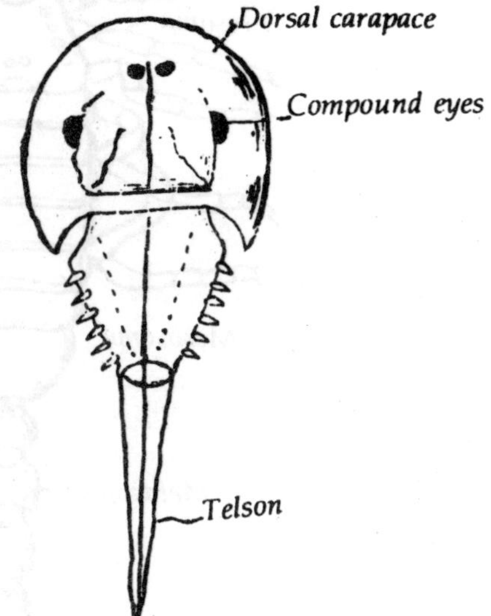

Limules

Class ARACHNIDA

1. The members comprises both aquatic and terrestrial forms such as *spiders, scorpions, mites and ticks.*
2. *Cephalothorax in this group shows 2 chelicerae, 2 pedipalps and 4 pairs of walking legs.*
3. Opisthosoma or abdomen is without legs.
4. Simple eyes are present.
5. Excretion is by coxal glands or malpighian tubules.
6. Respiration is by gills, trachea or book lungs.
7. They are dioecious and mostly oviparous. Some are Viviparous.
8. Courtship is seen before mating.

Examples

Aranea, Palamnaeus, Buthus, Ixodes.

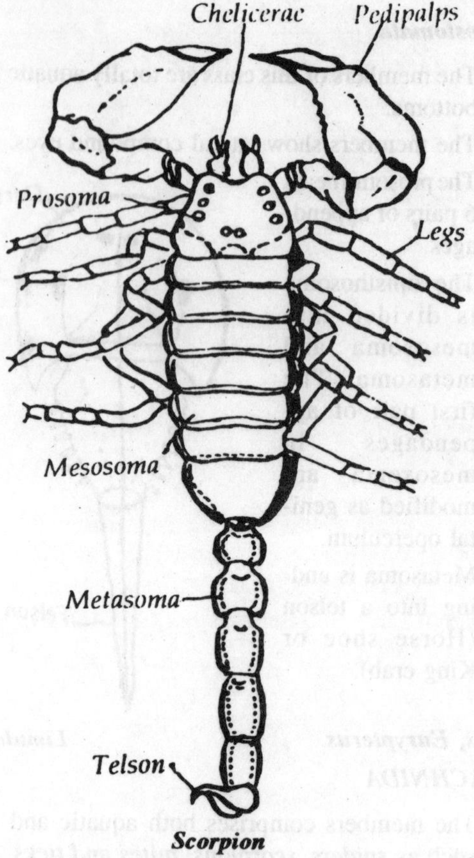

Scorpion

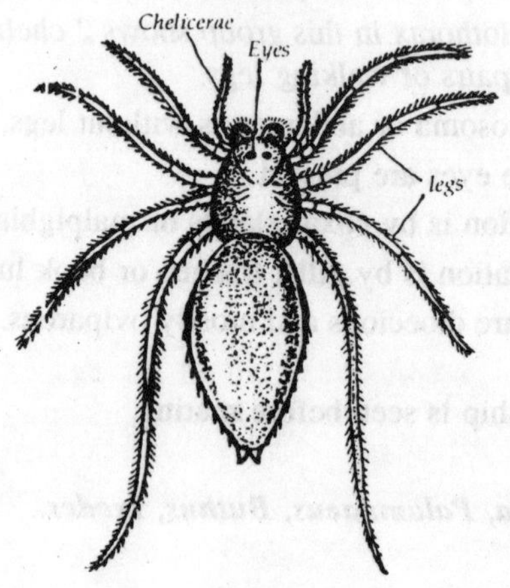

Spider

Sub phylum MANDIBULATA

1. The body of the members of subphylum Mandibulata (L., mandibula = mandible, ata = group) is divided into head, thorax and abdomen.
2. The head appendages are constituted by one or two pairs of antennae, a pair of jaws or mandibles and one or two pairs of maxillae.
3. Compound eyes are present in majority.

The subphylum mandibulata is divided into 4 classes namely, Crustacea, Diplopoda, Chilopoda and Insecta.

Class CRUSTACEA

1. The species of the class Crustacea are the major group of aquatic arthropods. They are predominantly marine, but there are fresh water species. A few have been invaded into the terrestrial habitat.
2. Crustaceans are unique among arthropods in having two pairs of antennae. The other characteristic appendages are a pair of mandibles and two pairs of maxillae.
3. The appendages in Crustacea are typically biramous.
4. Gills except in very small species are present. They are associated with the appendages. The number, location and shape of these gills vary greatly in different crustacean species.
5. Excretory organs are a pair of the green glands. These open at the base of the second antennae or the second pair of maxillae.
6. The sense organs in crustacea include a pair of compound eyes and a small median dorsal nauplius eye. The nauplius eye is characteristic of many crustacean larvae, but does not persist in the adult of many groups.
7. Copulation is typical in several crustaceans, egg brooding is very common. In the larval history, the first stage to be seen is the nauplius larva.

Examples

Streptocephalus, Daphnia, Mysis, Palaemon.

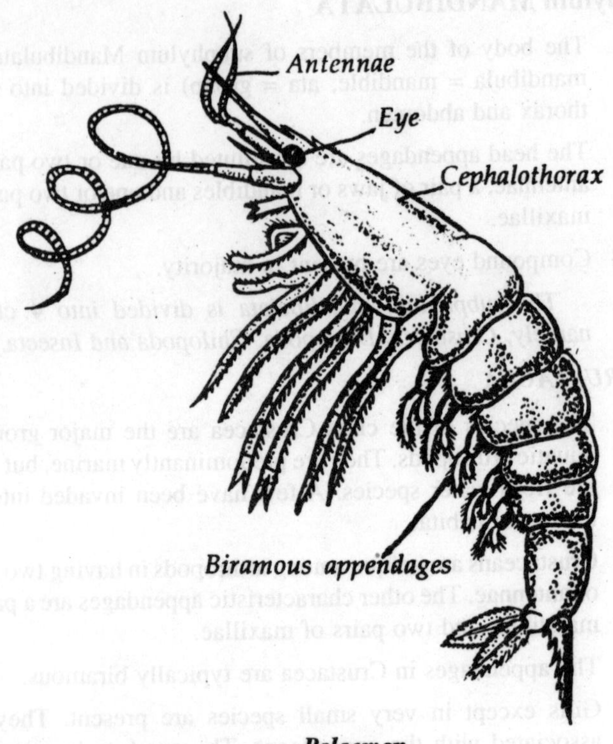

Palaemon

Class DIPLOPODA

1. This group includes millipedes or thousand-leggers.
2. These are nocturnal and herbivorus arthropods.
3. Two groups of simple eyes are present.
4. Body is elongated and cylindrical. It is divided into head, thorax and abdomen. The head bears 5 segments, the thorax bears 4 segments and the abdomen has 20 to 100 segments.
5. Thorax bears one pair of legs on each segment.
6. Trunk segment is a diplosegment. It bears a pair of legs.
7. Head bears one pair of antennae, one pair of mandibles and one pair of maxillae.
8. Genital openings are located midventrally on the third segment.

Examples

Julus, Polydesmus, Oxidus, Spirostriptus, etc.

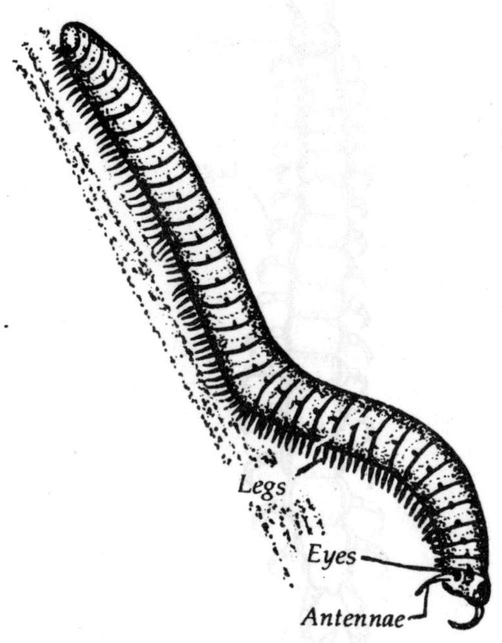

Spirostreptus

Class CHILOPODA

1. The members are commonly called as *centipedes* or *hundred leggers.*
2. Body is elongated and flattened dorsoventrally. It is divisible into head and trunk.
3. Head bears one pair of antennae, one pair of mandibles and two pairs of maxillae.
4. Trunk has 2 segments and each segment bears a pair of legs.
5. *First pair of legs are modified as poisonous claws.*
6. Genital opening is midventral on last but one segment.
7. These are nocturnal and carnivorus organisms.

Examples

Scolopendra, Scutigera, Lithobius.

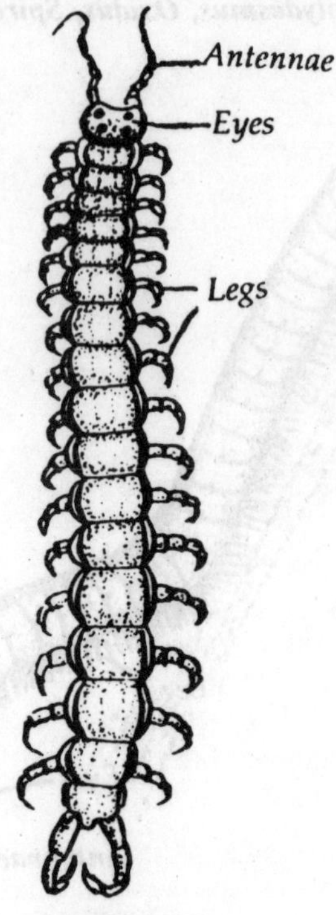

Scolopendra

Class INSECTA

1. Body is dividible into head, thorax and abdomen.
2. Head is constituted by 6 fused segments. Thorax is formed by 3 segments and abdomen is having not more than 11 segments.
3. Head bears a pair of compound eyes. It has paired appendages like antennae, mandible and maxillae.
4. The mouth parts in insects are modified according to their food habits. They are biting and chewing, piercing and sucking, sponging and sucking, siphoning, chewing and lapping types.
5. Thorax is with 3 pairs of legs and one or two pairs of wings.
6. Respiration is by trachea. They open to the outside by spiracles.

7. Sexes are separate, fertilization is internal, many insects exhibit metamorphosis.

Examples

Lepisma, Mautis, Pediculus, May flies, Stone flies.

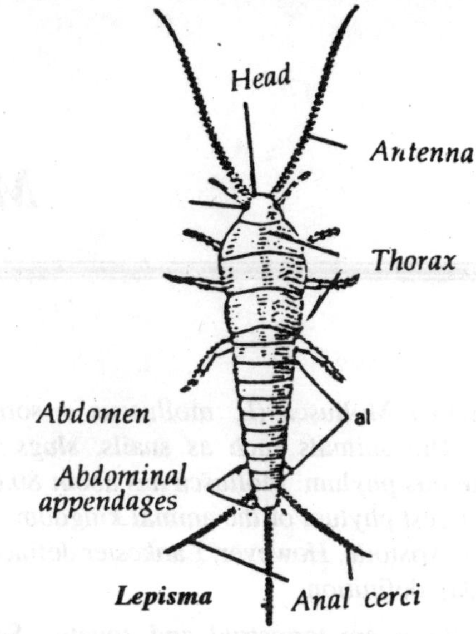

MOLLUSCA

Phylum Mollusca (L. molluscus = soft) includes soft bodied animals. The animals such as snails, slugs, clams, oysters, squids constitute this phylum. Mollusca has about 80,000 species and it is the second largest phylum of the animal kingdom. The term Mollusca was coined by Aristotle. However, Lankester defined the phylum true to the present day definition.

Molluscs are terrestrial and aquatic. Several molluscs live in marine waters and some live in fresh waters. They are free living animals, but a few Entoconcha, Thyonicola are parasitic.

The body of molluscs is covered by a mantle and protected by a shell. The shells are of various shapes, sizes and colours. The study of these molluscan shells is called Conchology, and the study of molluscs in general is called Malacology.

GENERAL CHARACTERS

1. The body in Mollusca is divisible into head, mantle, foot and visceral mass.
2. Molluscs are triploblastic, coelomate, unsegmented animals.
3. They show bilateral symmetry.
4. These animals show tissue grade of body organization. They are unsegmented, soft bodied animals.
5. The body in Mollusca is usually covered by an exoskeleton. The shell is made up of calcium carbonate. The shell is present

in a great majority of molluscs. It may be rudimentary or well developed. It is unívalved or bivalved. The shell in Mollusca is formed by the mantle which covers the body.

6. Appendages are absent in Mollusca.
7. Coelom is reduced. It is seen in the kidneys, pericardial cavity and gonadial cavity.
8. Digestive system is well developed in Mollusca. A rasping organ, the radula and a digestive gland, liver are present in the digestive system.
9. The respiration is direct. It is by gills or lungs or both.
10. Circulatory system is closed type, but the blood also flows through the sinuses. The blood contains a pigment called haemocyanin.
11. Excretion is carried out by paired metanephridia.
12. Nervous system is well developed. It is paired ganglia, connectives and nerves. The ganglia usually form a circum-enteric ring.
13. The sense organs in Mollusca are eyes, statocysts and receptors for touch, smell and taste.
14. Sexes are separate, a few molluscs are monoecious. One or two gonads with gonoducts, opening into the renal ducts or to exterior, are present.
15. Fertilization is external or internal. Development is direct or through larval forms, like trochophore or veliger larva.

CLASSIFICATION

Phylum Mollusca is divided into the following classes.

Class I : APLACOPHORA

1. These are primitive forms with worm-like body.
2. They are devoid of foot and shell, but mantle is present.
3. The head is differentiated from the body.
4. Calcareous spicules are seen burried in the cuticle.
5. Simple radula is present.
6. Eyes, statocysts and tentacles are absent.
7. A pair of gonads are present.

Examples

Neomenia, Neomematomenia, Chaetoderma.

Class II : POLYPLACOPHORA

1. These forms have dorsoventrally flattened body.
2. A small head which is not distinct from the rest of the body is seen.
3. Eyes and tentacles are absent.
4. Radula, mantle, foot and external gills are present.
5. Mantle cavity is seen posteriorly.
6. Shell is made up of 8 calcareous dorsal plates.
7. Gonad is single. A pair of nephridia are present.
8. Sexes are separate.
9. In the development, verliger larva is present.

Examples

Chaetopleura, Chiton.

Chiton

Class III : MONOPLACOPHORA

(Gr. monos = one + plax = plate + pherein = bearing)

1. These are marine forms.
2. Body is bilaterally symmetrical with a dome-shaped mantle.
3. The shell in these forms is flattened limpet-shaped with spirally coiled protoconch.
4. Foot is flat and broad with 8 pairs of pedal retractor muscles.
5. Five to six pairs of gills in pallial grooves are present.

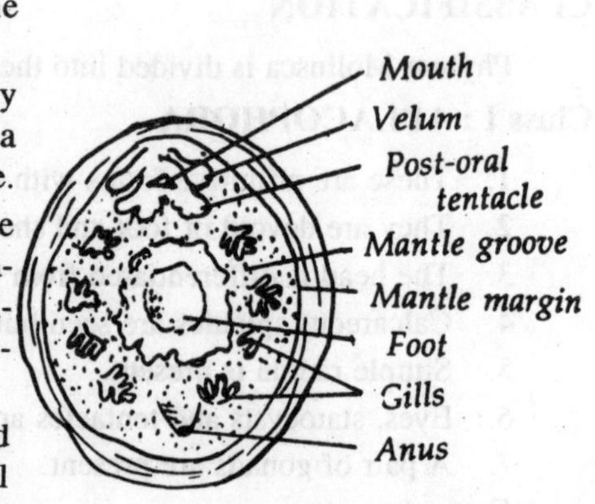

Neopilina

6. Six pairs of nephredia of which 2 are gonoducts are seen.
7. Radula is present in radular sac. Much coiled intestine is present.
8. Heart in these forms is constituted by 2 pairs of auricles and a single ventricle.
9. Nervous system is with longitudinal pallial and pedal cords.
10. These forms show internal segmentation.

Example

Neopilina.

Class IV : SCAPHOPHODA

(Gr. Scapha = boat + podos = foot)

1. These forms show tusk shells.
2. The body of these forms is seen with a tubular shell. The tube is open at both ends.
3. Head and eyes are absent. Mouth bears tentacles.
4. A conicle foot is present.
5. Radula is present.
6. Gills are absent.
7. Paired kidneys are present. Single gonad is seen.
8. These forms are dioecious. Trochophore larva is present in the embryonic development.
9. These are marine forms.

Examples

Dentalium, Pulsellum

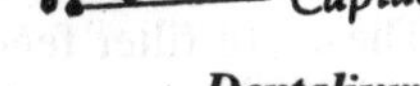

Dentalium

Class V : GASTROPODA

(Gr. gaster = belly + podos = foot)

1. This class includes snails and slugs. These are asymmetrical forms.
2. Shell is present or sometimes absent. When present, it is univalved and spirally coiled.
3. The torsion (coiling) of body mass is seen in the development of some of these forms.

4. Head is well developed.
5. Eyes, tentacles and radula are present.
6. A large flat foot is present.
7. A single kidney is present.
8. Veligar larva is seen during development.
9. These forms occur in marine, fresh water and terrestrial environments.

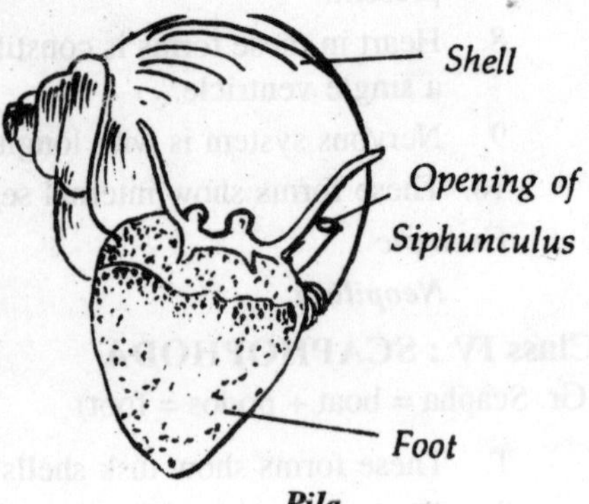

Pila

Examples

Pila, Patella, Aplysia, Murex.

Class VI : PELECYPODA

(= **Bivalvia**, Lamellibranchia)

(Gr. pelekus = hatchet + podos = foot)

1. These forms have laterally compressed body. This body is enclosed in a bivalve shell.
2. Head, tentacles, eyes, jaws and radula are absent.
3. A hatchet-shaped foot is present.
4. The alimentary canal is highly coiled and present in the foot.
5. The heart shows a single ventricle and two lateral auricles.
6. Statocysts and Osphraediaare present.
7. These are filter feeding forms.
8. They are usually dioecious. Veliger or glochidium larva occurs in development.
9. Majority of these forms occur in marine environment, but a few are fresh water inhabitants.

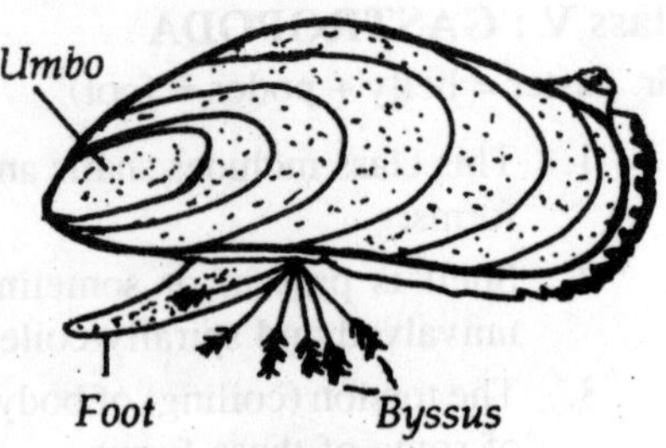

Mytilus

Examples

Mytilus, Ostrea, Lamellidens.

Class VII : CEPHALOPODA
(= Siphonopoda)

(Gr. cephale = head + podos = foot)

1. These forms are bilaterally symmetrical.
2. Body is dorsoventrally elongated.
3. The shell is external, internal or absent.
4. The head is distinct and large with well developed eyes. The foot encircles the head and is divided into a number of tentacles or pedal arms and siphon.
5. Radula is present.
6. Ink gland and duct are present in some forms.
7. These are dioecious forms. Development is direct.
8. These are marine free swimming forms.

Examples

Sepia, Nautilus, Loligo, Octopus.

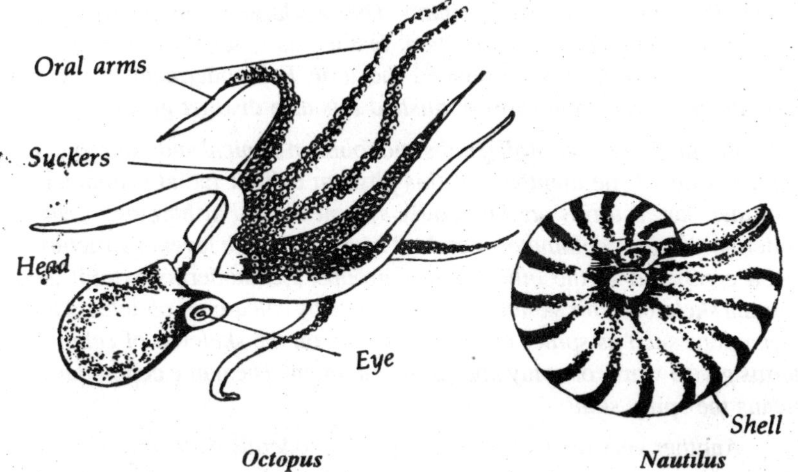

Octopus ***Nautilus***

ECHINODERMATA

The phylum Echinodermata (Gr. echinos = hedgehog + derma = skin + ata = characterised by) contains 6,000 species and it is the only major group of deuterostome invertebrates. The members of the phylum Echinodermata are marine inhabitants. They are largely bottom dwellers. They include star fishes, brittle stars, feather stars, sea cucumbers, sea lilies, etc. Jacob Kleins introduced the term Echinodermata to these animals, however, Leuckart established this as a distinct group.

Echinoderms live at all depths and found in tropical and subtropical marine waters. The members of this phylum exhibit radial symmetry. But, their larval forms are bilaterally symmetrical. Furthermore, echinoderms are true coelomate animals. They have a high level of structure when compared to the other radiate groups. Echinoderms exhibit an internal skeleton. This skeleton is constituted by calcareous ossicles. The projections such as spines or tubercles seen on the skeleton of echinoderms give a warty or spiny appearance to them. The name echinoderm means the spiny skin.

Another very distinct character of echinoderms is the presence of a unique system which of coelomic canals and surface appendages forming a water vascular system takes up the function of collecting and transporting food. It is also locomotor in function in majority of the echinoderms. The group Echinodermata possesses a spacious coelom in which a well developed digestive tract is present. Excretory system is absent in echinoderms. The gaseous exchange system vary from one group to another. The members of the phylum Echinodermata are commonly dioecious. Their reproductive tracts are very simple. Copulation is absent and external fertilization is seen in these animals.

Echinoderms are used as food in certain parts of the world.

GENERAL CHARACTERS

1. Echinoderms are triploblastic, coelomate and radially symmetrical animals with pentamerous disposition.
2. Echinoderms are marine animals with organ systems. Their body is not segmented. They exhibit various shapes such as globular, star-like, elongated, sperical, etc.
3. Head is absent in the echinoderms.
4. Body surface has fine symmetrically radiating areas called ambulacra and five alternating inter-radiary called inter-ambulacra.
5. The exoskeleton covering the body is constituted by calcarious ossicles or plates.
6. The water vascular system in this group is coelomic in origin. It is formed by podia or tube feet and madreporite. The tube feet aid in locomotion.
7. Coelom is of enterocoelous type. Coelomic fluid is with coelomocytes.
8. Vascular system referred to as haemal system is poorly developed. It is an open type of lacunar system. Heart is absent.
9. Digestive tract is simple. The alimentary canal is straight or coiled. Anus is present or absent.
10. Respiratory organs are dermal branchiae or papulae, tube feet, respiratory tree and bursae.
11. Nervous system is poorly developed. It is without a brain, but with a circum-oral ring and radial nerves. Sense organs are poorly developed. They include tactile organs, chemoreceptors, photoreceptors and statocysts.
12. Sexes are separate in echinoderms (dioecious). However, sexual dimorphism is not seen. Gonads are single or multiple. They are large in size. External fertilization is seen. Development is indirect and free swimming larval stages are seen in the development.
13. Echinoderms exhibit high power of regeneration. Autotomy or self-mutilation is seen in this group.

CLASSIFICATION

Phylum Echinodermata is divided into two subphyla. They are : **Eleutherozoa** (Gr. eleutherosoa = free + zoois = animal) and **pelmatozoa** (Gr. Pelmatos = stalk + zoois = animal).

I. Subphylum ELEUTHEROZOA

1. These are free living echinoderms.
2. A stalk is absent.

This subphylum is divided into four classes. They are 1. Asteroidea, 2. Ophiuroidea, 3. Echinoidea and 4. Holothuroidea.

Class 1. ASTEROIDEA

(Gr. aster = star + eidos = form)

1. This class includes star fishes or sea stars.
2. They have usually fine radiating arms with a central disc. The disc and arms are not sharply demarcated.
3. The coelom and its contents project into the hollow arms. Ambulacral grooves are open.
4. The anus and madreporite are situated aborally.
5. The tube feet with suckers and pedicellariae are present.
6. Bipinnaria larva is seen in the life history.
7. These are free living, slow creeping, predacious, scavengerous echinoderms.

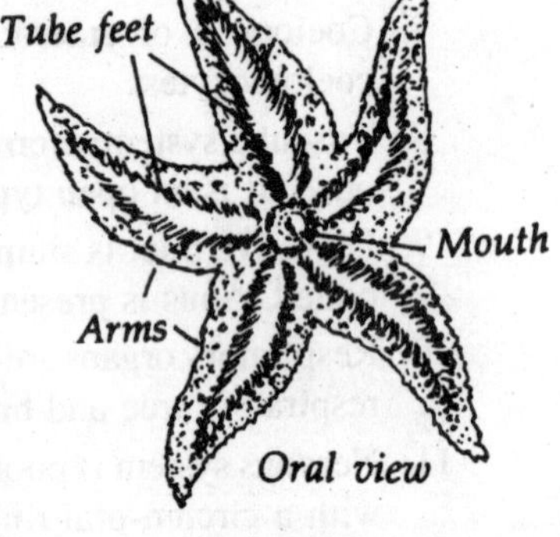

Asterias

Examples

Asterias, Astropcten, Heliaster.

Class 2. OPHIUROIDEA

(Gr. ophis = snake + oura = tail + eidos = form)

1. These echinoderms are star-like with sharply marked arms from the central disc.
2. The coelom and its contents are not extended into the narrow arms.

3. Pedicellariae are absent.
4. Sac-like stomach is present. Anus is absent.
5. The ambulacral grooves are absent or covered by ossicles. Tube feet is without suckers.
6. Madreporite is present on the oral side.
7. Ophiopluteus larva is seen in the life history.

Examples

Ophiura, Ophithrix, Asteronyx, Gorgonocephalus.

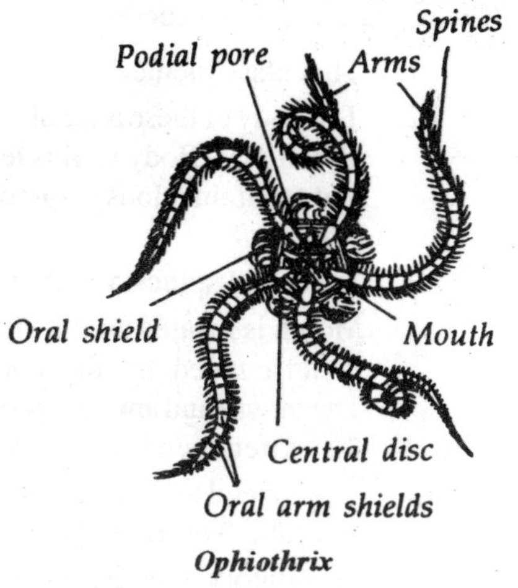

Ophiothrix

Class 3. ECHINOIDEA

(Gr. echinos = hedgehog + eidos = form)

1. This class includes sea urchins and dollars.
2. Their body is discoid, oval or semi-spherical. Arms are absent. There are five radial ambulacral areas alternating with five ambulacral zones.
3. Skeleton or test is compact bearing movable spines and three jawed pedicellariae.
4. Aristotle's lantern or chewing apparatus with teeth is present.
5. Ambulacral grooves are covered by ossicles.
6. Tube feet with suckers are present.
7. Anus and madreporite are present on the aboral side.
8. Gonads are usually five or less.
9. Echinopluteus larva is seen in the life history.

Examples

Echinus, Diadema, Clypeaster, Hemipneuster, Stomopneuster.

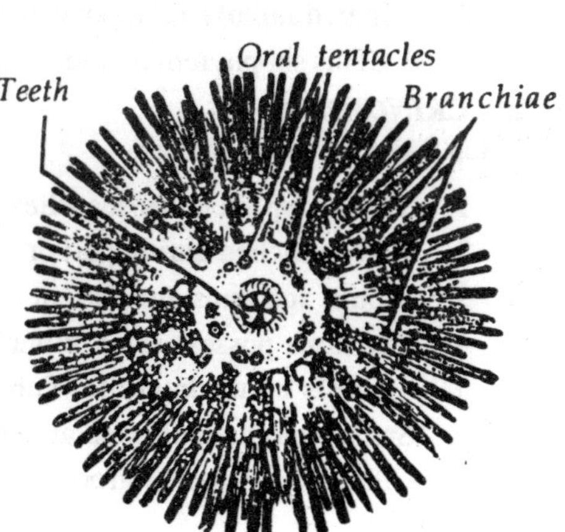

Echinus

Class 4. HOLOTHUROIDEA

(Gr. holothurion = sea cucumber + eidos = form)

1. This class includes sea cucumbers.
2. The body of these animals is cylindrical and elongated on oral-aboral axis. Body wall is leathery and contains loose spicules or plates.
3. Arms and spines are absent.
4. Mouth is located anteriorly and it is surrounded by the tentacles. The mouth and anus are present at the extreme ends of the body.
5. Ambulacral grooves are concealed and tube feet are with suckers.
6. Respiratory trees are present near cloaca.
7. Auricularia larva is seen in the life history.

Cucumaria

Examples

Holothuria Cucumaria, Molpadia, Synapta.

Subphylum II. PELMATAZOA

1. This subphylum includes sedentary echinoderms.
2. These forms are stalked.
3. Ciliated ambulacral grooves acting as food grooves are present.

This subphylum includes a single class, namely Crinoidea.

Class CRINOIDEA

(Crinon = lily + eidos = form)

1. This group includes sea lilies.
2. Body attached during part or whole of life by an aboral stalk.
3. Mouth and anus are present on the oral side.
4. The arms possess small pinnules.
5. There is no madreporite. Tube feet are without suckers.
6. Spines and pedicillariae are absent.
7. Ciliated ambulacral grooves are present on the oral side.

8. Pentacrinoid larva is seen in the life history.

Examples

Antedon, Neometra.

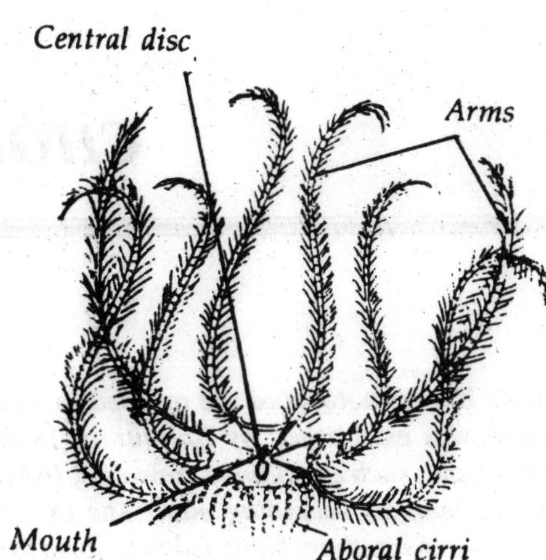

Antedon

CHORDATA

All animals having notochord are referred to as Chordates. The phylum Chordata was established by ***Balfour*** *in 1880. This phylum includes 45,000 species such as* ***acidians, lancelets, fishes, amphibians, reptiles, birds and mamals including man****. The chordates exhibit an astonishing diversity of structure, habit and functional aspects. However diverse they may be, they possess certain common characters. From this, it is evident that the chordates were derived from a common ancestral stock. There are different opinions about the origin of chordates.* ***Garstrong was of the opinion that chordates are evolved from the Auricularia larva of Holothuroidea of Echinodermata****. The ancestral chordates have appeared in the early part of* ***Palaeozoic era.***

GENERAL CHARACTERS

The chordates are closely related to the echinoderms and hemichordates. It is evident by the following resemblances. 1. The phosphogens used in muscle contraction are similar in hemichordates and echinoderms. The phosphogen is in the form of **creatin** in all vertebrates and **arginin in invertebrates.** In hemichordates and echinoderms, both creatin and arginin are present. 2. The tornaria larva of hemichordates resembles the larval forms of echinoderms. *So, it is believed that the ancestral chordates might have evolved from the larvae of echinoderms by paedogenesis or neoteny.*

Balfour included cephalochordates, unrochordates and vertebrates in Chordata. **Bateson** included Hemichordata in this phylum. Now-adays, the group Hemichordata is treated as a separate phylum. *All vertebrates are chordates but all chordates are not vertebrates...* The chordates show many common characters. There are three distinct characters of Chordata. They are—1. **Notochord, 2. Nerve Cord** and **3. Gills slits.**

1. Notochord : It is a stiff elastic mid-dorsal longitudinal rod found between the nerve cord and alimentary canal. *So, it is called chordadorsalis.* It is a skeletal or supporting structure. It is made of vacuolated cells and enclosed by two connective tissue sheaths. It persists throughout the life in some animals like ***Amphioxus, Lampetra*** and ***Petromyzon.*** In the larva of tunicates, the notochord is present in tail only. Notochord is invariably present in the embryos of the vertebrates including man, but replaced by the vertebral column or back bone in adult condition (it is the fore-runner of back bone). Notochord serves as an organ of internal support and also provides surface for the attachment of muscles. *The notochord is mesodermal in origin. In mammals, notochord is in the form of vestigeal swellings in the vertebral column. They are called Nuclei pulposi.*

2. Dorsal Tubular Nerve Cord : *The nerve cord in chordates is a single hollow tube, situated dorsal to both the alimentary canal and notochord.* During the development, it is formed on the dorsal side of the early embryo from ectodermal cells. In higher chordates, the nerve cord is differentiated into anterior brain and posterior spinal cord. *The nerve cord gets degenerated during metamorphosis in ascidians.*

3. Gill Slits : These are paired perforations on each side of the body towards anterior end leading form pharynx to the exterior. Gills are situated in *pharyngeal slits or branchial clefts.* Gill slits may be retained throughout the life and help in respiration or they may be present only during embryonic life. *In fishes and some adult amphibians,* each gill slit is lined with slender filaments, profusely supplied by blood vessels. These structures form the gill, meant for respiration. *In reptiles, birds and mammals several pairs of non-functional gill slits are formed only during embryonic period.*

4. Bilateral Symmetry : Various organs of the body are symmetrically arranged on either side of the central longitudinal axis. In a bilaterally symmetrical animal, the body can be divided into two equal halves which may be mirror images of each other.

5. Triploblastic Condition : *Various organs in the chordate body are derived from three germ layers, namely ectoderm, endoderm and mesoderm. Thus, all chordates are triploblastic animals.*

6. Coelom : *A true coelom or body cavity which develops from mesoderm is present in all chordates. But it is called enterocoelus coelom in chordates, as it develops from archenteron.* The coelom is externally lined by parietal or somatic layer and internally by visceral or splanchnic layer. In higher chordates like mammals, the coelom is divided into anterior thoracic and posterior abdominal cavities by a diaphragm.

7. Metamerism (*Segmentation*) : *The linear repetition of the body organs is called metamerism. Among the chordates, metamerism is seen mainly in the internal structures. The segmentation in chordates is referred to as heteronomous segmentation.* The metamerism is visible only internally in higher vertebrates. The body musculature is primitively segmental and many of the internal systems such as the nervous system, the circulatory system, the excretory system, etc., are in origin atleast, thoroughly segmental.

8. Endo and Exoskeleton : All vertebrates possess an unique type of skeleton not found in any invertebrate. *The internal skeleton is either cartilogenous or bony, derived from mesoderm. Exoskeleton in chordates is in the form of scales, dermal plates, feather, hairs, hooves, horns, nails, claws, etc.*

9. Cephalization : All the higher chordates possess well developed head having complex brain and specialized sense organs. This trend towards the prominence of head is called cephalization.

10. Ventral heart : The heart is muscular and contractile organ which is always situated towards ventral side below the alimentary canal. The heart is surrounded by double walled pericardium. It is 2 to 4 chambered in various groups of vertebrates.

11. Circulatory System : Blood flows from anterior to posterior end in dorsal blood vessel, whereas in invertebrates the flow of blood is from behind to forward in dorsal blood vessel. *As the blood flows through a system of tubes, the blood vascular system is of closed type.*

12. Hepatic portal system : *Hepatic portal system is present in all the chordates.* Blood that is collected from various parts of alimentary canal is not carried directly to heart but to liver. The blood then goes to heart.

13. Red Blood Corpuscles : In higher chordates, the respiratory pigment haemoglobin is always found in the specialised cells called red blood corpuscles or erythrocytes.

14. Post anal tail : *This is usually present in majority of chordates. In man and apes, it is well marked in the embryos, but is vestigial in adults.* In many aquatic and terrestrial vertebrates, it serves as an important locomotory organ. Tail is the posterior prolongation of the body without coelom and visceral organs.

15. Paired appendages : Higher chordates possess two paired appendages in the form of fins or limbs.

16. Endocrinal glands : Endocrinal glands present in vertebrates secrete **hormones.** They pour their secretions directly into blood. These glands are also known as **ductless glands.**

17. Blastopore : An embryonic pore which develops into anus is present in chordates. *Thus, all chordates are referred to as deuterostomous animals.*

CLASSIFICATION

Phylum Chordata is divided into four subphyla. They are :

1. Hemichordata 2. Urochordata 3. Cephalochordata 4. Vertebrata or Craniata

*Of these four subphyla, the first three are generally referred to as protochordates. All protochordates are characterised by the absence of cranium, so they are also called **acraniates.** The protochordates stand between invertebrates and vertebrates.*

Subphylum 1 : HEMICHORDATA

(Adelochordata)

1. Hemichordata includes marine, worm like animals.
2. *There is no notochord, but a stomochord or buccal diverticulum which arises from the wall of the buccal cavity is present.* This stomochord differs from notochord in its position, structure and mode of development. Hence, *hemichordates are called doubtful chordates.*
3. Paired gill slits are present on the dorsolateral sides of the trunk region.
4. The body is divided into three regions, namely, proboscis, collar and trunk.
5. *There is no brain and these are unsegmented animals.*
6. Both dorsal and ventral nerve cords are present.

7. The excretory organ is in the form of glomerulus.
8. The flow of blood in blood vessels is of invertebrate type.
9. Development is indirect or direct. A tornaria larva which resembles echinoderm larvae occurs in the development.

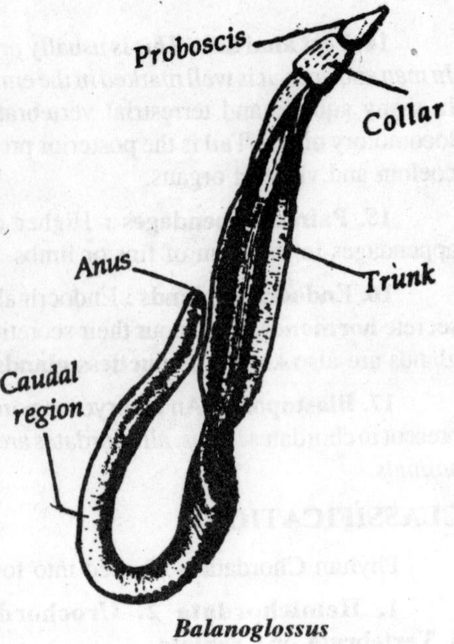

Balanoglossus

Examples

Balanoglossus, Tachyglossus, Cephalodiscus, Rhabdopleura.

Subphylum 2 : UROCHORDATA

1. Exclusively marine. Some are sedentary and some are pelagic forms.
2. *These are degenerated chordates. Their body is enveloped by a tunic or test made by tunicin (cellulose like material). So they are called tunicates.*
3. In the larval form, nototchord is present only in the tail region.

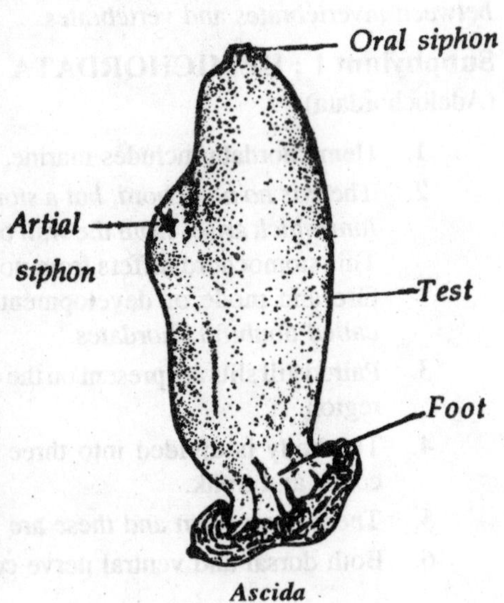

Ascida

4. Nerve cord is found only in the larval stage, but absent in adults.
5. *The larval form is called Tadpole and it exhibits retrogressive **metamorphosis.***
6. These urochordatess are generally referred to as **ascidians**.

Examples

Ascidia, Doliolum, Salpa, Pyrosoma, Oikopleura.

Subphylum 3 : CEPHALOCHORDATA

1. *These are small fish-like transparent, marine animals.*
2. Pharynx is perforated by numerous gill slits.
3. Nerve cord is tubular and dorsal.
4. Segmentation and coelom are well developed.
5. ***Notochord is extended from the posterior to the anterior end.***
6. Nephridia are the excretory organs.
7. Myotomes, nephridia, gonads and gill slits are metamerically arranged.

Examples

Amphioxus, Asymmetron.

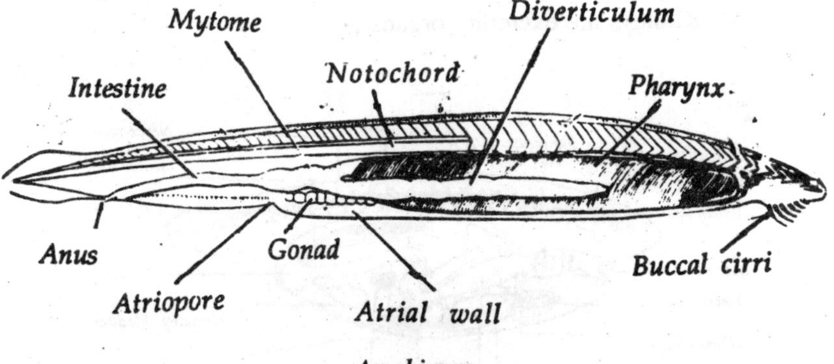

Amphioxus

Sub-phylum 4 : VERTEBRATA (Craniata)

1. These are highly evolved, complex chordates with distinct heads.
2. *Notochord is partially or totally replaced a vertebral column. Such animals are called vertebrates.*

3. The anterior part of the nerve cord is enlarged to form the brain.
4. ***A cranium or brain box encloses and protects the brain.***
5. Visceral clefts are not more than 7 pairs.
6. Two pairs of appendages which help in movements are present in the form of fins and limbs.
7. Endoskeleton is cartilagenous or bony.
8. *Lungs are the respiratory organs except in fishes and some amphibians.*
9. Respiratory pigment is haemoglobin which is found in erythrocytes.
10. *The heart is 2 chambered in fishes, 3 chambered in amphibians and 4 chambered in birds and mammals. In reptiles, incompletely divided 4 chambered heart is present.*
11. Hepatic portal system is present in all vertebrates. *Renal portal system is present in all vertebrates except mammals.*
12. Liver and pancreas are found in association with digestive system.
13. Endocrinal glands which pour their secretions into blood are present in all vertebrates.
14. Kidneys are excretory organs.

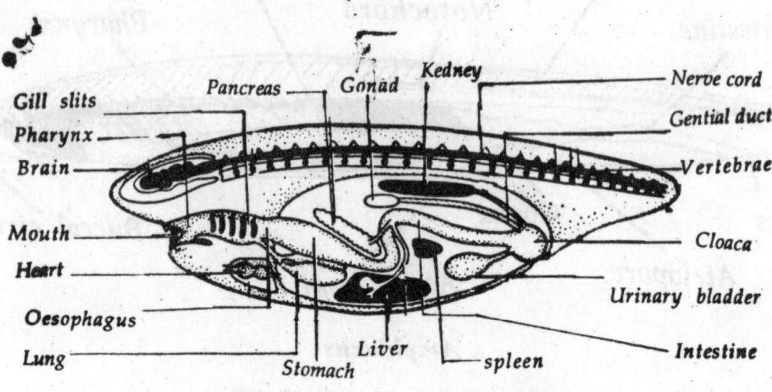

V.S. of Vertebrate

Subphylum Vertebrata is divided into 2 superclasses viz.,

1. Agnatha and **2. Gnathostomata**

Super class 1. AGNATHA

1. This class is characterised by the earliest vertebrates without jaws and paired appendages.
2. A median single nostril is present.

Superclass Agnatha is divided into 2 classes, viz., 1. *Ostracodermi* and 2. *Cyclostomata.*

Class 1 : OSTRACODERMI

These are the most ***primitive, oldest*** and ***extinct jawless vertebrates*** with a bony armature.

Examples

Cephlaspis, Hemicyclospis, Clearaspis.

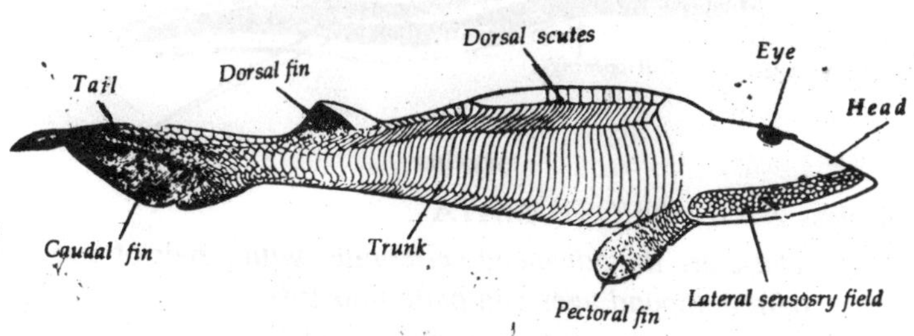

Hemicyclospis

Class 2 : CYCLOSTOMATA

1. *These are living fish-like jawless vertebrates.*
2. Paired fins and scales are not found.
3. Endoskeleton is cartilagenous.
4. Heart is 2 chambered.
5. *Mouth is round and suctorial.*
6. *Notochord is persistant throughout the life.*
7. A single median nostril is present.

8. 6–14 pairs of gill slits are present.
9. *These lead parasitic life on fishes.*
10. Development may be direct or indirect. *The larval form in lampreys is called Ammocoetus which resembles Amphioxus.*

Examples

Petromyzon, Myxine

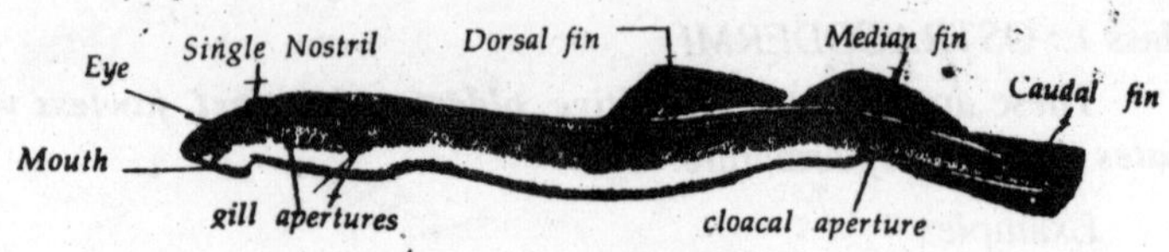

Petromyzon

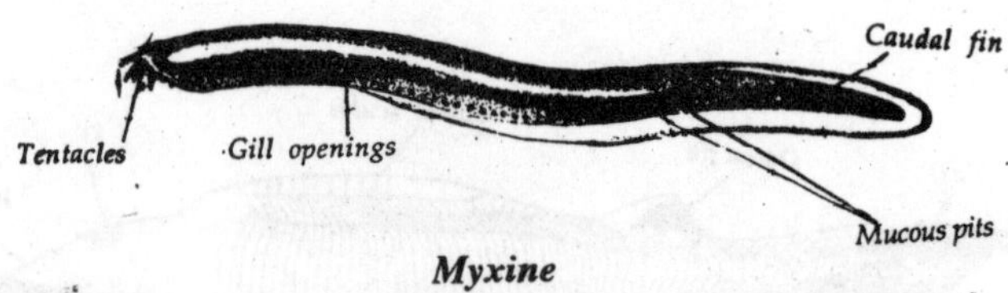

Myxine

Super class 2 : GNATHOSTOMATA

1. *These are the advanced vertebrates with a pair of true jaws, paired append ages and paired nostrils.*
2. The endoskeleton is well developed.

This superclass is divided into 7 classes. These are :

1. Placodermi, 2. Chondrichthyes, 3. Osteichthyes, 4. Amphibia, 5. Reptilia, 6. Aves, and **7. Mammalia.**

The animals of the classes Placodermi, Chondirichthyes and Osteichtyes are collectively called Pisces. The animals of the remaining four classes are known as tetrapods because they have 2 pairs of pentadactyle limbs. In fishes and amphibians, the extra **embryonic or foetal membranes** are not formed during development, hence they are called **Anamniota.**

In the development of **reptiles, birds and mammals**, an extraembryonic membrane viz. amnion is formed hence they are called **Amniota.**

Pisces, amphibians and reptiles are **cold blooded** or **poikilothermic** or **ectothermal** vertebrates which cannot maintain constant body temperature. **Aves** and **mammals** are **warm blooded** or **homeothermic** or **endothermal vertebrates** as they have thermoregulatory mechanizm. Pisces and amphibians are included in the group ***Ichthyopsida,*** where as reptiles and aves are grouped into ***Sauropsida.***

PISCES

Pisces are the first gnathostome (jawed) vertebrates and the first successful class of chordates. They are world wide in distribution. Pisces (fishes) are aquatic, gill breathing, cold blooded vertebrates. A fish has a streamlined body so as to facilitate easy movement in water. ***There are about 26,000 species of fishes and they represent the largest single group of vertebrates.*** They exhibit such variations in their body size. The smallest fish is ***Mystichthys lozerenses*** and the largest fish in Rhinodon sp.

Fishes are evolved during the Ordovician period and they dominated during ***Devonian period.*** Hence, Devonian period is popularly known as "***Golden age of fishes***". Among fishes some are extinct (placoderms) and a few (Coelacanth) remain unchanged from the early ages.

Amphibians have evolved from fishes (Crossopterygian stock) by the end of Devonian period.

GENERAL CHARACTERS

1. *Fishes are aquatic, ectothermal vertebrates.*
2. The body is spindle-shaped.
3. *The body is covered by various types of dermal scales. They are mesodermal in origin.*
4. Respiration is effected by four or five pairs of gill slits which are lined by gills. In some fishes such as Dipnoi, a hollow pharyngeal outgrowth called *air bladder* functions as a lung.

5. Slime glands are abundantly found in the body wall. Their secretion keeps the skin slimy and reduces the friction with water.
6. *The locomotion is effected by special appendages called fins. Besides locomotion, they also help in balancing.* **Pectoral** and **pelvic fins** are paired. **Dorsal ventral fins** are unpaired in nature. **The caudal fin functions as a rudder.**
7. One pair of external nostrils is present on the anterodorsal surface of the head. They do not communicate with the mouth cavity except in Dipnoi fishes. These nostrils are useful in respiration in dipnoi. In other fishes, they function as olfactory organs.
8. Neck and tympanum are absent. Eyes are without eye lids. There is a nictitating membrane moving back and forth over the eye ball.
9. The endoskeleton may be cartilaginous or bony. The notochord is replaced by vertebrae.
10. *Only a single occipital condyle is present in the skull, hence the fish skull is monocondylic.*
11. Vertebrae are of amphicoelous type.
12. External and middle ears are absent, only internal ear is present. Internal ear has three semicircular canals.
13. Alimentary canal is much more specialised with a definite stomach and with well developed liver and pancreas. The alimentary canal ends into anus or cloaca.
14. *The 'S' shaped heart (Venous and single circuit heart) has one auricle and one ventricle and possesses only the venous blood except in Dipnoi where the auricle is partially divided into two parts. Sinus venosus, conus arteriosus an bulbus arteriosus are found in association with heart.*
15. Red blood corpuscles are oval, biconvex and nucleated.
16. Both hepatic and renal portal systems are present.
17. Brain is small and it does not fill the cranial cavity.
18. Cranial nerves are ten pairs.
19. *Lateral line sense organs are well developed and situated on the head and along the trunk. These organs help to detect the pressure changes in water (rheoreceptors).*

20. The functional adult kidney is always a **mesonephros**. There is no urinary bladder.
21. The nitrogenous excretory waste is in the form of **ammonia** in bony fishes and **urea** in elasmobranchs.
22. Sexes are separate, gonads are typically paired and the genital ducts open into cloaca.
23. Fertilization is either external or internal.
24. Extra-embryonic membranes are absent.
25. Most of the fishes are oviparous and a few fishes are viviparous (Scoliodon).
26. **Eggs are large with much yolk (megalecithal).**
27. Development is generally direct without any metamorphosis.

The group Pisces can be divided into three classes.

I. *Placodermi*

II. *Chondrichthyes*

III. *Osteichthyes*

Class I : PLACODERMI

1. *These are extinct fishes and first formed jawed vertebrates.*
2. They lived during Silurian, Devonian and Carboniferous periods of **Paleozoic era**.
3. The placoderms have heavy bony plates on their bodies.

Examples

Climatius, Bothriolepis, Palaeospondylus.

Climaticus

Class II : ELASMOBRANCHII OR CHONDRICHTHYES.

1. *These are cartilagenous fishes and includes sharks, skates and rays.*
2. *They are exclusively marine.*

3. A spiracle is present in front of the first gill slit.
4. Mouth and nostrils are ventral in position. Teeth are covered by enamel-like substance called **vitrodentine.**
5. 5-7 pairs of gills are present. Operculum is absent.
6. Body is covered by minute placoid scales or dermal denticles.
7. Intestine has a spiral valve (shark). **Alimentary** canal opens out by cloaca.
8. Tail heteroeercal.
9. They possess urea in the blood. It is an adaptation to live in highly saline water.
10. ***Males are provided with claspers*** *(modifications of pelvic fins) for copulation. These claspers are present on either side of the cloaca.*
11. Fertilization is internal.
12. Some are viviparous.

Examples

Scolidon, Torpedo, Narcine, Stegostoma, Sphyrna, Pristis.

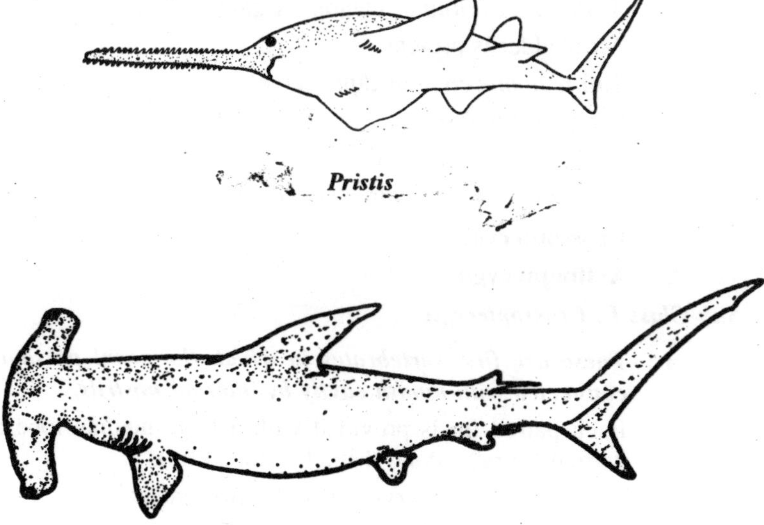

Pristis

Hammer head shark

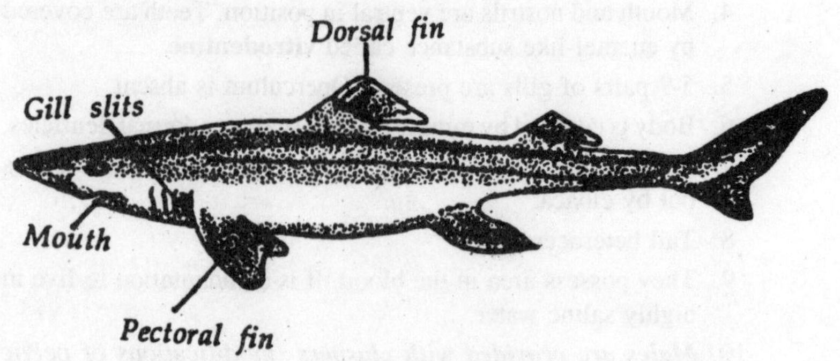

Scoliodon

Class II : OSTEICHTHYES

1. This includes bony fishes.
2. ***Endoskeleton is bony in nature.***
3. Mouth is usually terminal.
4. *These fishes are found both in marine and fresh waters.*
5. *Four pairs of gills are present and they are covered by a bony operculum.*
6. *Scales are cycloid or ctenoid or ganoid.*
7. Air bladder is present.
8. *Tail is homocercal or diphycercal.*
9. Claspers are absent.
10. Fertilization is external.

Osteichthyes is divided into sub classes.

1. **Crossopterygii.**
2. **Actinopterygii**

Sub-Class I : Crossopterygii

1. ***These are first vertebrates in which the nasal passage is connected with mouth cavity by internal nostrils.***
2. Each paired fin is provided with a large median lobe and dermal fin rays along the sides (lobed fin).

The sub-class Crossopterygii includes two orders:

1. *Rhipidistia*
2. *Dipnoi*

Order 1: Rhipidistia

It includes *extinct osteolepids* and *extant coelocanths*. These are believed to be extinct during the **Mesozoic era.** *However, a coelacanth fish was discovered in living condition in 1938 along the coast of South Africa. This fish was named after a curator,* ***Miss. Latimer*** *as* ***latimeria.*** It is steel blue in colour measuring 5 ft. in length and weighing 127 lbs. Since it has not undergone any change in millions of years, it is regarded as a ***'living fossil'***

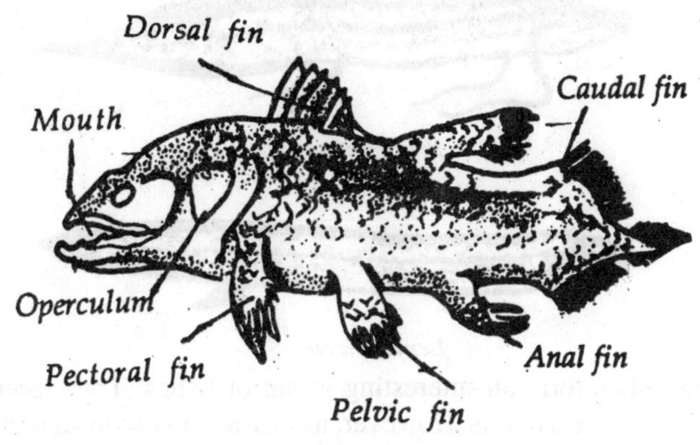

Latimeria

Order 2: Dipnoi

It includes lung fishes which are amphibious and found in the tropical rivers of southern hemisphere. The body is long and slender and covered by *cycloid scales.* Caudal fin is *diphycercal.* Internal nares are present. Air bladder in these fishes is modified into lung. Endo and exo-skeletons are reduced. Endo-skeleton is bony.

Lung fishes were evolved during the **Devonian** and flourished during the **Permian** and **Traiassic periods**. They are now represented by only three genera. The three living genera are distantly distributed over three continents, thus they are said to be exhibiting ***discontinuous distribution.***

The three representative genera are:

1. ***Neoceratodus:*** It is found in the Burnet and Marry rivers of **Queens-land of Ausrialia.**
2. ***Protopterus :*** It is found in the rivers of Senegal, White Nile and Lake Tanganiyaka of **Africa.**

3. ***Lepidosiren :*** It is found in the river **Amazon of South America**.

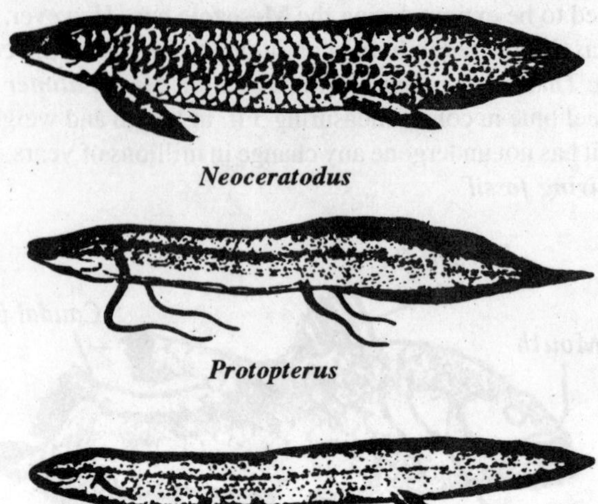

Neoceratodus

Protopterus

Lepidosiren

Lung fishes form an interesting group of fishes. They resemble amphibians in several points. However, a close affinity with amphibians is not established. According to a modern view, Dipnoi and Amphibia are possibly derived from some common stock. *According to **Romer**, an eminent paleontologist, Dipnoi are not the ancestors, but **'uncles'** of Amphibia.*

Sub-class: ANTINOPTERYGII

Antinopterygii (ray fined fishes) includes three super-orders.

Super-order 1: Chondrostei

Examples

Acipenser (Sturgeon) (North America, Europe)

Super-order 2: Holostei

Example

Amia (Bowfin) (North and Central America).

Super-order 3: Teleostei

It includes a large number of modern bony fishes. They occur both in marine and fresh waters.

Examples

Exocoetus, Hippocampus, Echeneis, Cynoglossus.

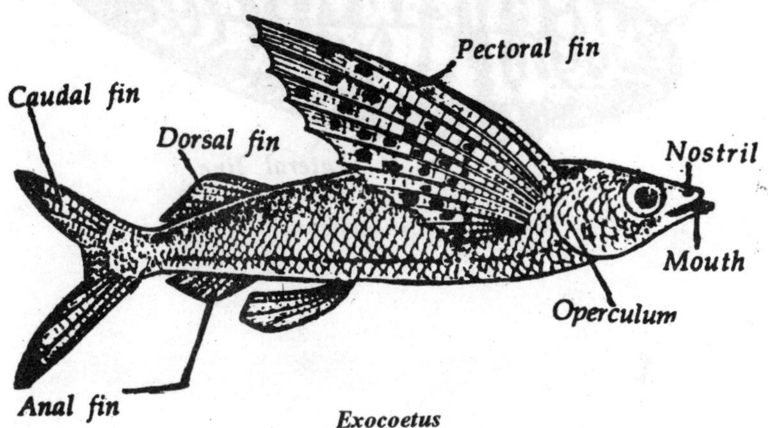

Exocoetus

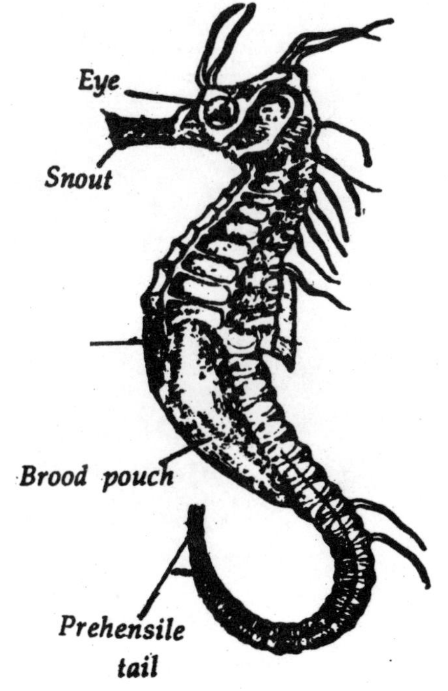

Hippocampus

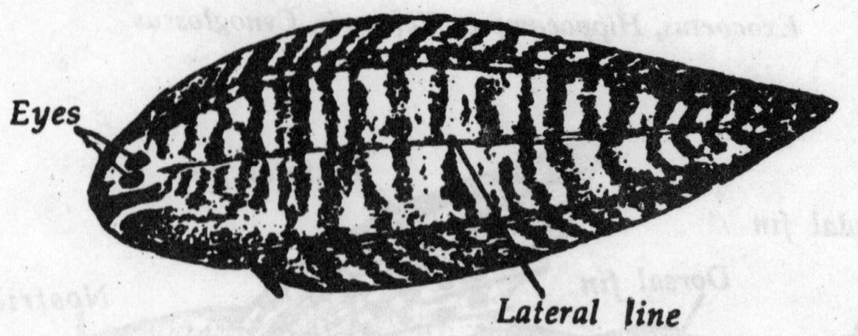

Cynoglossus

AMPHIBIA

The amphibians are vertebrates which lead a dual mode of life. Most of the amphibians live partly in fresh water and partly on moist soil. ***Amphibians are the first and primitive tetrapods*** *(tetra = four, pod = foot).*

The earliest amphibians came into existence during the Devonian period. The extinct osteolepids were the closest relatives of Amphibia. The striking similarities between the extinct osteoplepids and the earliest fossil amphibians such as ***Ichthyostega, Eryops*** confirmed the origin of Amphibia from the Crossopterygian stock. ***The lobed find of Osteolepids is the fore-runner of pentadactyle limb of Tetrapoda.***

Amphibians stand between the fishes and reptiles both in structure and function representing the transition of life from water to land. A number of changes had taken place in the organization of the body of an amphibian. The development of pentadactyle limb, internal nares, moist scaleless skin, tympanum, pulmonary respiration etc., allowed the earliest amphibians to come out of water and to live for sometime on land. However, the amphibians could not emerge as a successful group on land due to their inability to complete their development on land. This drawback prevented the amphibians from becoming completely terrestrial.

The amphibians were predominent during the Carboniferous period. Hence, the Carboniferous period is regarded as the age of Amphibia. Most of the amphibians became extinct by the turn of ***Mesozoic era*** leaving behind a few representatives as modern Amphibia. Thus, the class Amphibia includes both extinct and extant (living) forms.

*The living amphibians are represented by **frogs** and **toads** (tailless forms) and caecilians (limbless forms). **Amphibians are worldwide in distribution. However, they are not found in marine habitat.***

Amphibians show a lot of variation in their size and shape. The toads and caecilians are mostly terrestrial, while frogs and urodeles are mostly aquatic.

*The class Amphibia consists of about **250 living species**. They show a number of common characters.*

GENERAL CHARACTERS

1. Amphibians are tetrapods which live partly in water and partly on moist soil.
2. *The body temperature in amphibians changes with the changes in the environment. Hence, they are called poikilothermous or ectothermal or cold blooded tetrapods.*
3. The body of amphibians is divisible into head, trunk and tail. But, tail is absent in frogs and toads. Neck is absent in all forms.
4. The skin in Amphibia is glandular and moist. Scales are generally absent except in caecilians. *The moist scaleless skin serves as an organ of respiration.*
5. *The body of amphibians generally bears two pairs of pentadactyle limbs. But, they are absent in caecilians (Apoda). The limbs are of equal size in Urodela and unequal size in Anura.*
6. The external nostrils communicate with the buccal cavity through a pair of internal nostrils. The development of internal nostrils paved the way for pulmonary respiration.
7. *A pair of eyes with immovable upper eye lid and movable lower eye lid is located on the head.* The lower eye lid bears a transparent **nicititating membrane** (third eye lid). Eyes are degenerated in caecilians and in some urodeles.
8. *There is no external ear in Amphibia. **Tympanum** is present only in frogs and toads. **Columella auris** is the only ear ossicle present in the middle ear, it is a modified **hyomandibular.***
9. **Salivary glands** are absent in Amphibia.
10. Maxillary teeth are usually present. But teeth are totally absent in toads.
11. There is a sticky protruding tongue which helps in capturing food.

12. The endoskeleton is mostly bony in nature.
13. *Skull in Amphibia bears two occipital condyles. Hence, it is described as dicondylic.*
14. Sternum is first developed in Amphibia. It is absent in caecilians.
15. *Vertebrae in Amphibia are procoelous or amphicoelous type.*
16. Ribs are usually absent when the ribs are present in some forms, they do not reach the sternum.
17. Digestive system is well developed in Amphibia. The alimentary canal opens into cloaca at the posterior end.
18. There is no distinct tracheal tube due to the absence of neck *The voice box in Amphibia is laryngotracheal chamber.*
19. *A pair of elastic, bladder-like lungs is present to carry on pulmonary respiraticn.*
20. *The red blood cells are oval, biconvex and nucleated in amphibians.*
21. *The heart in Amphibia is 3 chambered with 2 auricles and 1 ventricle.* Truncus arteriosus and sinus venosus are also associated with the heart.
22. Both hepatic and renal portal systems are present in the venous system.
23. The cerebellum of brain in Amphibia is poorly developed and the olfactory lobes are indistinct.
24. *There are 10 pairs of cranial nerves in Amphibia as in the case of fishes.*
25. There are no lateral line sense organs in adult amphibians.
26. *The excretory organs are mesonephric kidneys. Nephrostomes are associated with the kidneys in Amphibia.*
27. Adult amphibians excrete urea. Hence they are known as **ureotelic animals.**
28. Amphibians are generally ***oviparous***. Eggs are ***mesolecithal*** (eggs with moderate amount of yolk).
29. *Cleavage is holoblastic and unequal.*
30. Copulatory organ is generally absent. But it is present in caecilians.
31. Development is usually indirect. The larva undergoes metamorphosis to become adult.

32. Parental care is seen in many amphibians.
33. *Romer described amphibians as defeated group, as they are not able to come to land completely.*

CLASSIFICATION

The class Amphibia consists of both extinct and extant forms. The extinct amphibians are included in three orders. The are as follows :

Order 1 : Labyrinthodontia

These are crocodile-like fossil amphibians of Carboniferous period. *The salient features of Labyrinthodontia are a skull with complete bony roof and larged teeth with infolded enamel. The Labyrinthodontia gave rise to cotylosaurs.*

Examples

Eryops, Cocopus.

Order 2 : Phyllospondylia

These are small fossil amphibians with large heads, short bodies and feeble limbs. *The Phyllospondylia gave rise to Anura and Urodel.*

Example

Branchiosaurus.

Order 3 : Lepospondylia

These are small extinct amphibians with ribs. *The Lepospondylia gave rise to Apoda.*

Examples

Lysorophus, Diplocaulus.

The extant Amphibia is divided into 3 orders. They are as follows:

Order 1 : Apoda or Gymnophiana

They are limbless amphibians commonly known as caecilians or blind worms or naked snakes.

The characters of caecilians are :

1. Body of caecilians is elongated and worm-like. The skin is wrinkled with small cycloid scales embedded in the wrinkles or grooves. Adapted for burrowing life.
2. Eyes are degenerated and covered by the skin.
3. *The head bears a single sensory tentacle. It is present between nostril and eye.*

4. Tympanum, limbs and girdles are absent.
5. Notochord is persistent.
6. *Vertebrae are amphicoelous.* Ribs are present, but they do not reach the sternum.
7. *A ductus botalli joining the systemic artery and pulmonary artery is present. Lungs asymmetrical.*
8. The clocal wall of male is eversible and serves as **copulatory organ**. *Presence of copulatory organ is unique among amphibians.*
9. Fertilization is internal.
10. Some females lay eggs in strings in the soil and coils around the strings of eggs till they are hatched. Thus, the female caecilian show parental care.

Three genera of caecilians are found in Kerala and Karnataka regions of South India.

Examples

Ichthyophis, Uraeotyphlus, Gegenophis

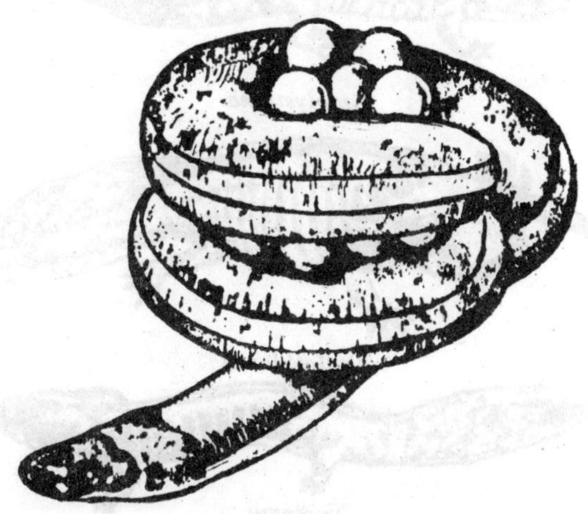

Ichthyophis

Order 2 ; Urodela or Caudata

These are scaleless amphibians with a tail in the adult stage. Urodela comprises of ***salamanders*** and ***newts***. Urodeles are discontinous in distribution. They are mostly found in **North America** which is recognised as the *head quarters of Urodela.*

The Characters of Urodela are :

1. *Urodeles have two pairs of short weak limbs which are of almost equal size.*
2. Eyes are small and degenerated in most of the urodeles.
3. *Persistent external gills are found in most of the adult urodeles.*
4. Teeth are found in both jaws.
5. Urodeles are usually oviparous.
6. The larvae are aquatic and resemble adults in several species. ***The larva of Amblystoma tigrinum*** (tiger salamander) is ***Axolotl***. The Axolotl larva remains in larval stage for a prolonged period and attains sexual maturity. **This process is known as neoteny or paedogenesis.**

Examples

Necturus, Proteus, Siren, Salamandar, Amblystoma, Tylototriton.

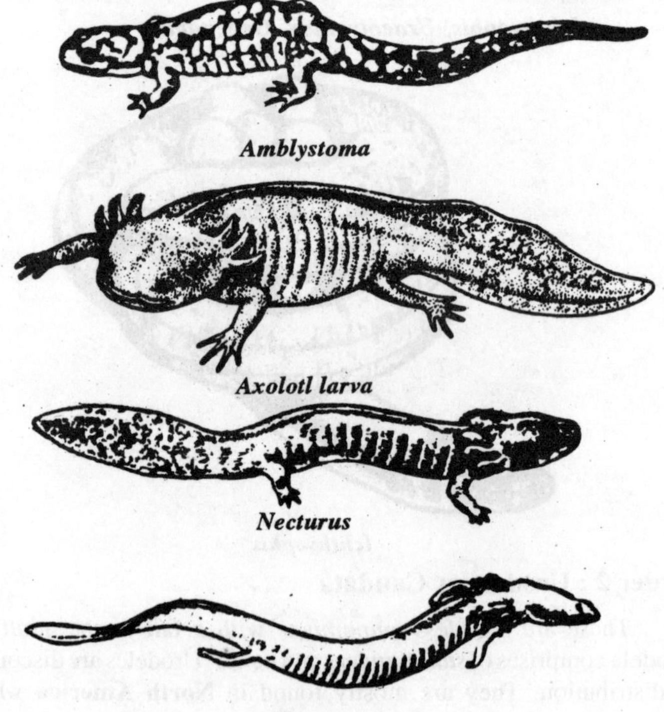

Amblystoma

Axolotl larva

Necturus

Proteus

Order 3 : Anura or Salientia

These are scaleless amphibians without a tail in adult stage. Anura is represented by **frogs** and **toads** which are cosmopolitan in distribution.

The characters of **Anura** are :

1. Body is divisible into head and trunk.
2. Vertebrae are less in number (5–10) Elogated urostyle is present.
3. *Fore limbs are short while hind limbs are long.*
4. Tympanum is present.
5. *Well developed eyes with monocular vision are found both in frogs and toads.*
6. Gills are absent in the adult stage, but they are present in larval stage.
7. Fertilization is external.
8. *The larval stage of Salientia is known as a Tadpole.*

Examples

1. Rana hexadactyla 2. Bufo melanosticus 3. Rhacophorus.

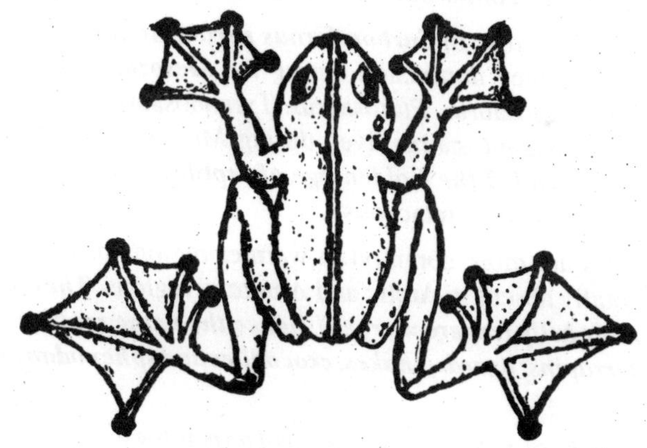

Rhacophorus

REPTILIA

Reptiles are the first vertebrates which become adapted completely for terrestrial life.

They are advanced than amphibians. The reptiles lay eggs on land. The embryo is protected by embryonic membranes. The development in reptiles takes place outside the aquatic medium.

Reptiles are ectothermal and lung breathing vertebrates.

*The reptiles arose in **Carboniferous period** from **Labyrinthodont amphibians** and they dominated during the **Mesozoic era**. The giant dinosaurs like Plesosaurus, Pterosaurus, Tyrannosaurus, Ichthyosaurus, Brontosaurus, Giagantosauras lived during Mesozoic era. **Hence, the Mesozoic era is called the Golden age of reptiles.** Most of the reptiles became extinct in late Cretaceous.*

*Reptiles are more common in warmer regions of the world and some are totally absent in Arctic and Antarctic regions. **They generally creep on their belly (repere = creep), hence the name reptiles. Turtles, tortoises, terrapins, lizards, snakes, crocodiles and sphenodon are living reptiles.***

The study of reptiles is known as Herpetology.

GENERAL CHARACTERS

1. *Reptiles are cold blooded or poikilothermous or ectothermal vertebrates i.e. the body temerature is variable according to the environmental changes.*

2. The skin is dry and skin glands are absent.
3. *Body is covered with horny epidermal scales or scutes. This is an adaptation for terrestrail mode of life.*
4. Body is divided into four regions namely head, neck, trunk and tail. Two pairs of pentadactyle limbs are usually present. Limbs are modified into paddles in aquatic reptiles, but are lost secondarily in snakes.
5. The ear consists of **middle** and **internal ears** (middle ear is absent in snakes).
6. *Skeleton is completely ossified. Skull is with a single occipital condyle (monocondylic); true sternum is present except in snakes and ribs are attached to sternum; vertebrae are procoelous; vetebral column is divided into cervical, thoracic lumber, sacral and caudal regions; the temporal region of skull bears temporal fossae and arcades; tibia-fibula are free; the pectoral girdle has an interclavicle.*
7. *Heart is imperfectly four chambered with two auricles and partially divided ventricle (in crocodiles the ventricle is completely divided), arotic arches (systematic arches) are one pair, red blood corpuscles are biconvex, oval and nucleated; both hepatic and renal protal systems are well developed.*
8. *Respiration is effected by lungs (pulmonary) and lungs are spongy; cloacal respiration is found in some turtles; functional gills are never present.*
9. Teeth are found in both the jaws (except in Chelonia).
10. *The rectum is divided into three parts namely coprodaeum, urodaeum and proctodaeum.*
11. Excretion takes place by one pair of **metanephric kidneys;** *kidneys lack* **nephrostomes**; nitrogenous excretory waste is chiefly in the **form of solid uric acid.**
12. *Tweleve pairs of cranial nerves are present* (in snakes there are only 10 pairs).
13. The embryo during development is protected by extra-embryonic membranes. (amnion, allontois, chorion and yolk sac).
14. All are unisexual : males possess **copulatory organs** (except *Sphenodon*).
15. *Fertilization is internal; mostly oviparous* (except some snakes and lizards); *eggs are large megalecithal with abundant yolk*;

eggs are laid on land and they are protected by calcareous shell (cleidoic). As the yolk is situated towards one side of the egg, the reptilian eggs are described as **telolecithal.**

16. Cleavage or segmentation is **meroblastic** (incomplete or partial).
17. **Metamorphosis is absent** i.e. the young ones resemble the adults (development is direct).

CLASSIFICATION

The living reptiles belong to **four orders.**

Order 1 : Chelonia

1. *They are primitive living reptiles.* It includes marine turtles, terrestrial tortoises and fresh water terrapins.
2. *Limbs are paddle-like in turtles. Some times they bear claws.*
3. Trunk is enclosed in a bony shell composed of *dorsal carapace* and *ventral plastron* and the body can be retracted inside the shell.
4. *Temporal fossae are absent in the skull (anapsid skull).*
5. The jaws are *devoid of teeth (edentulous).*
6. Single **nasal opening** is present.
7. The pulmonary and systemic trunks on each side are connected by **ductus botalli.**
8. Cloacal aperture is longitudinal. **Copulatory organ is single.**
9. All are oviparous.
10. *All are herbivorous except some marine turtles.*

Examples

Dermochelis, Chelone mydas (Green turtle) Testudo (Terrestrial) Trionyx (fresh water).

Chelone mydas

Order 2 : Rhyncocephalia

(Rhynos = beak; cephalon = head)

1. *The only living representative of this order is Spenodon or Hatteria.* It is found only in New Zealand. It is also known as Tautara lizard.
2. It looks like a lizard. The skin is scaly and sales on the mid-dorsal line form a crest. The skin is olive green.
3. *Skull is diapsid and vertebrae are amphicoelous.*
4. *Proatlas is present between skull and Atlas in crocodiles.*
5. *There is a well developed penial or parietal eye on the head and it is sensitive to light.*
6. Sternum and abdominal ribs are also present; ribs are single headed and they bear uncinate processes.
7. *Copulatory organ* is absent. Urinary bladder present.
8. Clocal opening is transverse.
9. *Sphenodon is considered as living fossil.*
10. It is an *insectivorous, carnivorous* and *nocturnal reptile.*

Order 3 : Crocodilia

1. It consists of ***Crocodiles,*** and ***alligators.***
2. *These are the largest living fresh water reptiles.*
3. Tails is long and *laterally compressed,* which helps in swimming.
4. *Body is covered by bony plates (scutes) which are overlapped by horny epidermal plates.*
5. Teeth are thecodont and jaws are elongated.
6. *The skull is of typical diapsid type* (with two temporal fossae); there is a false palate in the skull.
7. Thoracic ribs possess uniciate processes as in birds.
8. *Vertebrae are procoelous. Pro-altas is present.*
9. *Heart is four chambered and the ventricle is completely divided into two by a septum leaving small foramen between. It is called foramen of punizza.*
10. *The thoracic cavity is separated from the abdominal cavity rated from the abdominal cavity by a membranous stransverse partition, called the diaphragm.*
11. Urinary bladder is absent.

12. Clocal aperature is longitudinal.
13. Single copulatory organ is present.

Examples

Crocodilus, Alligator, Gavialis.

Gavialis

Order 4 : Squamata

It is the largest order of living reptiles. It consists of lizards and snakes.

1. The body is covered with *horny epidermal scales.*
2. Temporal fossae in the skull become secondarily reduced to one pair in lizards and they are absent in snakes.
3. *Vertberae are procoelous.*
4. **Teeth are pleurodont** i.e., the teeth are fused with the outer surface of the jaw bones.
5. **Cloacal opening is transverse.** Males possess a *pair of eversible copulatory organs (hemipenis).*
6. *Ductus caroticus is present (a small blood vessel joining systemic and carotid arteries).*

This group includes two sub orders.

Sub-Order 1 : Lacertilia

1. It includes lizards and these are the most successful reptiles.
2. In majority, both fore and hind limbs are present, bearing clawed digits (Ophiosaurus and Barkudia are limbless).
3. **Eye lids are movable.**
4. **Tympanum is present.**
5. *There is a sternum and it has a T-shaped episternum (interclavicle).*

6. Two rami of the mandible are united.
7. Quadrate is immovable.

Examples

Varanus, Calotes, Hemidactylus, Draco, Chamacleon, Heloderma.

Chamaeleon

Varanus

Sub-order 2 : Ophidia

1. ***Snakes*** belong to this group.
2. Body is covered with *scales.* Skin is moulted several times.
3. *Temporal fossae are absent secondarily.*
4. *Limbs, sternum, tympanum and urinary bladder are absent.*
5. The rami of mandibles are united by a ligament.
6. *Eye lids are immovable and tongue is bifid and protrusible.*
7. *Lung, kidney and gonad are asymmetrical* (left side organs are reudced).
8. About 2500 species of snakes are there.

Examples

Hydrophis, Naja naja, Bungarus fasciatus, Echis carinata, Python, Dryophis.

Naja naja

Bungarus fasciatus

Python

AVES

Aves are bipedal, warm blooded vertebrates with an exoskeleton of epidermal feathers and scales. Body temperature is very high and constant in birds. Birds resemble reptiles in many characters such as laying of eggs on land and formation of foetal membranes during embryonic development. Due to these common features, ***reptiles*** and ***birds*** included in the group ***Sauropsida***. Based on the resemblance between the fossil birds and the reptiles, Huxley described ***"birds as the glorified reptiles"***.

Aves were evolved during Jurassic period of Mesozoic era from ornithischian dinosaurs. Birds became modernized in the Cretaceous period. The modernization involved both morphological and physiological changes such as maintenance of constant temperature, fusion of skull bones, formation of pneumatic bones, fusion of vertebrae to form synsacrum. Birds have become highly modified to adapt to an aerial mode of life. ***There are about 8,590 species of birds which lead aerial, terrestrial and equatic modes of life.***

The study of birds is known as ***ornithology.***

GENERAL CHARACTERS

1. *Birds are warm-blooded (endothermal) vertebrates with an exo-skeleton of feathers.*
2. **The fore limbs are modified into wings.** Each wing bears three digits and provided with feathers for flight.
3. The hind limbs are adapted for walking, swimming and perching. They bear four toes.

4. The body is divided into four regions namely **head, neck, trunk** and **tail.**
5. Skin glands are absent, but an oil gland (preen gland or uropygeal gland) is present at the base of tail in flying birds.
6. *The body is covered by feathers except beak and legs.*
7. The jaws are protruded to form characteristic horny beak (rhamphothecka) and teeth are absent in extant birds.
8. The oesophagus gets dialated into **crop,** and the stomach is divided into **proventriculus** and **muscular gizzard. Cloaca** is three chambered. One pair of **rectal caecae** are found at the junction of the ileum and large intestine.
9. The bones, forming the skeleton, are spongy, light in weight containing air cavities (pneumatic bones).
10. **The skull bones are completely fused.**
11. The skull is sutureless and monocondylic (having single occipital condyle).
12. Cervical and free thoracic verterbrae are usually heterocoelous.
13. *Few thoracic, lumbar, sacral and anterior caudal vertebrae are fused to form a synsacrum.*
14. *The posterior caudal vertebrae are usually fused to form a pygostyle* (plough shaped bone).
15. Ribs are provided with uncinate processes.
16. The sternum is broad, usually with a logitudinal ventral keel or carina for the attachment of flight muscles.
17. *The coracoid is pillar-like, scapula is sabre-shaped.*
18. *Two clavicles are fused to form V-shaped furcula.*
19. *Distal carpals and metacarpals are fused to form carpo-metacarpus.*
20. Proximal tarsals are fused with the tibia to form a tibio-tarsus. Metacarpals are united with distal-tarsals to form tarso-meta-tarsus.
21. *Lungs are spongy and non-elastic. Air sacs are present in lungs* and some of them communicate with air cavities in the bones. Alveoli are absent in lungs.
22. *The voice is produced by the syrinx, situated at the junction of trachea and bronchi.* Larynx is without vocal cords.
23. The heart is four chambered, only the *right systemic arch* (aortic arch) is present in the adult.

24. *Erythrocytes are oval, biconvex and nucleated.*
25. ***Renal protal system*** is reduced.
26. *Kidneys are metanephric, three lobed, Urinary bladder is absent. Nitrogenous excretory waste is eliminated in the form of semi-solid uric acid.*
27. The cerebral hemispheres and cerebellum are large and the optic lobes are displaced laterally.
28. *Twelve pairs of cranial nerves are present.*
29. Sexes are separate and **sexual dimorphism** is well developed.
30. *Only left ovary and oviduct are present. Penis is absent in flying birds.*
31. Fertilization is internal. All are oviparous.
32. Eggs are **megalecithal type** containing much amount of yolk. Eggs undergo meroblastic cleavage.
33. Parental care is well developted.
34. *Extra embryonic membranes like amnion, allantois, chorion and a large yolk sac are formed during embryonic development.*

CLASSIFICATION

The class Aves is divided into two sub-classes :

1. Archaeornithes 2. Neornithes.

Sub-class 1 : Archaeornithes

1. *It includes two extinct Jurassic birds namely **Archaeopteryx** and **Archaeornis.***
2. These are oldest fossil birds.
3. The fossils of these birds are found in **Jurassic strata** of **Bravaria in Germany.**
4. They show characters of both reptiles and birds.
5. ***Archaeopteryx*** *is considered as the connecting link between reptiles and birds.*
6. ***Thecodont teeth*** are present in both the jaws.
7. *It is of the size of a crow and is an example for mosaic evolution.*

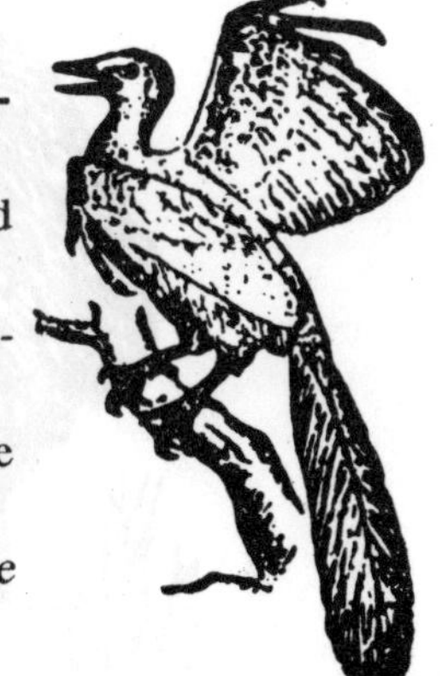

Archaeopteryx

Example

Archaeopteryx

Sub-class 2 : Neornithes

1. It includes both living and extinct birds.
2. Teeth are absent in modern birds.
3. Tail is usually short and it ends with a **pygostyle.**
4. The sternum is well developed and usually provided with a keel or carina.

This sub-class is divided into a three super-orders.

***Super-order 1 :* Odontognathae**

1. It includes extinct, flightless, aquatic **Cretaceous birds.**
2. Teeth present in furrows of jaws.
3. Sternum lacks keel.
4. Clavicles are not fused.

Examples

Icthyomis, Hesperornis.

***Super-order 2 :* Palaeognathae**

1. Ratitae or flightless birds are included in this group.
2. They are adapted for running or walking.
3. **Wings are reduced or absent.**
4. Teeth are absent, caudal vertebrae are free.
5. **Sternum is devoid of keel or carina.**

Examples

Ostrich, Emu, Kiwi, Tinamus

Kiwi

African Ostrich

MAMMALIA

Mammals (Gr : mamae = mammary glands) are highly evolved tetrapods and they are at the top of the animal kingdom. They were originated in Jurassic period about 180 million years ago from mammal-like reptiles, the therapsids (extinct cotylosaurian reptiles).

Various orders of mammals came into existence during the early part of ***Coenozoic era*** which is regarded as the ***"Golden age of mammals".*** Mammals are highly successful group of vertebrates. They are found in all parts of the world showing adaptive radiation.

The class Mammalia includes a variety of animals such as ***moles, rats, rabbits, bats cats, whales, horses, camels, elephants, apes, monkeys, man etc. Parental care is well developed in the class Mammalia.***

Study of mamals is known as Mammology.

GENERAL CHARACTERS

1. The body comprises *head, neck, trunk* and *tail.* The neck is prominent connecting the head with trunk.
2. *The body is usually covered with hair which is epidermal in origin.* Presence of hair is an important characteristic feature of mammals. Hair is sparsely distributed in aquatic mammals like whales. *Hair forms a sort of insulation around the body helping in the retention of body heat.*
3. The skin of mammals is thick and water proof. *The skin is provided with numerous glands like sweat, sebaceous, scent and mammary glands.* **Sweat glands (sudorific glands) are**

epidermal in origin but found in dermis. They regulate the body temperature and also act as excretory organs. **Sebaceous glands** open into hair follicles and they secrete an oily substance called sebum. The sebum keeps the skin soft and smooth. *The sebaceous glands in the pectoral region are modified into mammary glands except in prototherians.* The aquatic mammals have a thick layer of fat in the subcutaneous tissue. *This fat is known as blubber and serves as a thermal barrier.*

4. Mammals are tetrapods. **Two pairs of pentradactyl limbs,** variously adapted for walking, running, climbing, burrowing, swimming and flying are present. *Hind limbs are absent in whales and sea cows. Fore limbs are modified into flippers in aquatic mammals.*
5. The head bears a definite nose. Eyes are having movable eye lids with eye lashes to prevent the entry of dust particles into eye. *Nictitating membrane is vestigial in mammals (Pica semilunaris).*
6. **External ears** (auditory pinnae) are present. However, they are absent in aquatic *mammals, prototherians* and ant eaters.
7. *The middle ear contains a chain of bones called ear ossicles. These are malleus (hammer-like), incus (envil) and stapes (strirrup bone).* The internal ear has a long spirally coiled cochlea.
8. Teeth are firmly fixed in sockets of jaw bones and as such they are called **thecodont**. All the teeth are not similar and do not carry similar functions. They differ in shape and function. This condition is known as **heterodont dentition**. There are four kinds of teeth.
   a. **Incisors** are always born in front of the jaws and **serve for cutting.**
   b. **Canines** are long and sharp in carnivorous mammals and help in tearing the flesh. *The canines are absent in herbivourous mammals.*
   c. **Premolars** are *grinding teeth* with cusps.
   d. **Molars** (*wisdom teeth*) resemble premolars in their structure and function. The molars are not represented in milk dentition.

Premolars and molars are together described as cheek teeth. Generally two sets of teeth, the milk set (deciduous set) and a permanent set are formed in mammals. *This condition is referred to as diphyodont dentition. However, monophyodont dentition is seen in marsupials.*

9. *The buccal cavity is divided into an upper nasal passage and a lower food passage by a palate.* The respiratory aperture or glottis is guarded by a flap-like epiglotts. *There are 4 pairs of salivary glands in the buccal cavity of mammals. In man, there are 3 pairs of salivary glands.*
10. The body cavity is divided into an anterior thoracic cavity and posterior abdominal cavity by a *dome-shaped transverse muscular wall called diaphragm.*
11. The oesophagus is a long tube. It passes through neck and thorax. It joins the stomach below the diaphragm. There is a caecum at the union of small intestine and large intestine. *The caecum is long in herbivorous mammals like rabbit and it is short in carnivorous mammals like cat. The liver is five lobed (four lobed in man).* The anus is separated from urino-genital aperture. Cloaca is absent in mammals except in monotremes.
12. Respiration is effected by *large elastic lungs.* A Larynx with vocal cords is usually present. *The diaphragm plays an important role in breathing movements.*
13. *The heart is 4 chambered. Only left aortic (systemic) arch is present. Renal portal system is absent. The red blood corpuscles are round biconcave and enucleated.* Blood has blood platelets. Mammals are *homothermous animals.*
14. The skull has a large cranial cavity. The number of bones in the skull is reduced and they are firmly united by sutures. *The skull is dicondylic. The lower jaw is composed of a single bone called dentary on each side.*
15. *Mammals possess 7 cervical vertebrae with a few exceptions. The centrum with a few exceptions. The centrum is platycoelous.* Intervertebral discs are present between vertebrae. Vertebrae bear epiphyses at both ends of the centrum. The ribs are double headed. *Only scapula is well developed in pectoral girdle and the coracoid is very much reduced. An obturator foramen is present in the pelvic girdle.*

16. The brain is of the most advanced type. The surface of the **cerebrum possesses convolutions.** The two cerebral hemispheres are connected by a **transverse nerve band called corpus callosum**. There are **4 optic lobes** which constitute the corpora quadrigemina.
17. The cerebellum bears pons varoli.
18. Twelve pairs of cranial nerves are present.
19. *One pair of bean-shaped metanephric kidneys is present in the abdominal cavity* on either side of the backbone. A U-shaped Henle's loop is present in the nephron. *Nitrogenous waste in mammals is in the form of urea. Hence they are called* **ureotelic animals.**
20. The testes in mammals are situated outside the abdomen in scrotal sacs. **The inguinal canal joins the scrotal sac with the abdominal cavity**. Males possess a muscular, copulatory organ called penis.
21. Ovaries are situated in the posterior part of abdominal cavity. The oviduct consists of **an anterior coiled fallopian tube, a large median uterus and a posterior vagina.** A small broad clitoris is situated very close to the urino-genital aperture in female. **The clitoris is homologous to the penis of male.**
22. *Fertilization in mammals takes place in the fallopian tube.*
23. Mammals are viviparous. They give birth to young ones (except in Prototheria).
24. The eggs are usually minute without yolk. Hence, they are called **alecithal eggs.**
25. There is a definite period of intraiuterine development for the embryo. This period is known as the *period of gestation.* The gestation period varies from individual to individual.
26. *An allantoic placenta is present in the development of eutherian mammals. Foetal membranes like amnion, chorion* and *allantois* are formed in the development. The chorionic villi penetrate deep into maternal uterine wall to form a placenta. **The placenta supplies food and oxygen to the embryo** and collects then itrogenous waste from the foetal blood to be released into the maternal blood.
27. Mammals possess lips and cheeks.
28. ***Mammals are highly evolved and most intelligent of animals.***

CLASSIFICATION

The class Mammalia is divided into 3 sub classes viz., 1. *Prototheria*, 2. *Metatheria* or *Marsupialia* and 3. *Eutheria* or *Placentalia*.

Sub-class 1 : PROTOTHERIA

It includes the order Monotremata. Echidna and duckbill are the representatives of Monotremata. The Prototheria are primitive unfinished mammals. (Romer) prototherians possess a number of reptilian characters in addition to the mammalian features.

1. *The mammary glands of prototheria are modified sweat glands unlike the Eutheria.* Teatless mammary glands are characteristic of prototheria. *Males possess functional mammary glands. The feeding of young with milk by father and mother (parents) is known as* ***gynaecomastism.***
2. Auditory pinna or external ear is absent.
3. The skull is without tympanic bulla and lacrimals. **Vertebrae are without epiphyses.**
4. Cervical vertebrae bear ribs. Ribs are single headed having capitulum only.
5. *A large coracoid with precoracoid and a 'T' shaped interclavicle is present in pectoral girdle.*
6. The pelvic girdle has a pair of epipubic bones which are absent in eutherians.
7. *The body temperature is low and ranges from 25°C to 28°C. Hence, they are described as heterothermous mammals.*
8. A cloaca is present, into which ureters and urinogenital sinus open. Oviducts remain separate throughout the length and open into the cloaca.
9. Testes are abdominal. Penis helps to passout the sperms only.
10. *There is no corpus callosum in the brain.*
11. Teeth are present only in young ones; *adults lack teeth,* but a horny beak is present.
12. *Females are oviparous, laying large eggs with much amount of yolk.*
13. Optic lobes are 4 in number.

Reptilian features of Monotrems

1. A large coracoid with precoracoid is seen. T-shaped interclavicle in pectoral girdle is present.

2. Pelvic girdle is with epipubic bones.
3. Vertebrae are without epiphyses.
4. Ribs are single headed.
5. Cochlea is simple and less spirally coiled.
6. Auditory pinna is absent.
7. Cloaca is present and testes are abdominal.
8. Laying of large eggs with abundant yolk is seen.
9. Meroblastic cleavage is seen in the development.

Mammalian features of Monotrems

1. Hair is present all over the body.
2. Diaphragm is present.
3. Only left aortic arch is present.
4. Heart is 4 chambered.
5. Non-nucleated red blood corpuscles are seen in blood.
6. Mid brain shows four optic lobes.
7. Skin possesses mammary, sweat and sebaceous glands.
8. Three ear ossicles are present in the middle year.
9. A single bone (dentary) is present in each half of the lower jaw.
10. A slightly coiled cochlea is present.

This sub-class Prototherea includes 3 living genera. These are found in Australia, Tasmania and New Guinea.

Ornithorhynchus

1. ***Ornithorynchus*** or duck billed platypus : It is distributed in *Australia* and *Tasmania*.

2. ***Echidna*** or ***Tachyglossus*** or spiny ant eater : It is inhabited in *Tasmania* and *New Guinea.*

Echidna

3. ***Zaglossus*** or ***Proechidna*** : It is found in New Guinea and resembles *Tachyglossus.* Some authors grouped metatherians and eutherians collectively as Theria. There are some common features in both of these groups. They are :
    1. Ear is usually having an external pinna.
    2. Teeth are usually present in young and adult forms.
    3. Cloaca is usually absent.
    4. Testes are situated in scrotal sacs.
    5. Mammary glands are with teats.
    6. Females are viviparous.
    7. Oviducts open into the vagina.
    8. The distal part of vas deferens and urinary bladder join together to form a common passage, the urethra. The urethra opens at the tip of the penis.

Sub-class 2 : METATHERIA OR MARSUPIALIA OR DIDELPHIA

Marsupials (pouched mammals) are more advanced than the prototherians. *They are all pouched mammals characterised by the presence of an integumentary brood pouch or marsupium in which the immature young ones are fed with the milk of the mother.*

The marsupials occupy an intermediate position between the primitive mammals (monotremes) and the higher mammals (eutherians).

1. Marsupium or brood pouch is present in females. *The mammary glands are present in the marsupium.*

2. *The marsupium is supported by epipubic (marsupial) bones.*
3. The teeth are more in number. There are more than 3 incisors and more than 4 molars in each half of the jaw. Teeth are formed only once in life time. Hence, the dentition is called monophyodont dentition.
4. *Corpus callosum is usually absent in the brain.*
5. Placenta is absent.
6. Although the anal and urinogenital apertures are separate and distinct, they are surrounded and controlled by a common sphincter muscle.
7. Testes are situated in scrotal sacs. The penis is present behind the scrotum and it is bifid.
8. *The females have 2 oviducts, 2 uteri and 2 vaginae (didelphic condition) which separately open into urinogenital sinus.*
9. The young ones are born naked and blind, but *they posses clawed fore limbs* by which they move into the brood pouch.
10. Fertilization and a major part of development are internal, **but** the young ones are born in immature state (mammary **foetus**).
11. Yolk sac is large with villi.
12. *Hind limbs are long in some animals like kangaroo.*
13. The body temperature ranges from 36°C to 40°C.

The marsupials are now found only in certain 8 countries and they are good example for discontinuous distribution. All living marsupials are confined to Australia and it is described as the land of marsupials. In other parts of the world, they perished due to the arrival of carnivores. Australia was isolated from the main land long before the evolution of carnivorous eutherian mammals. Thus the island forms a safe home for the marsupials. Some marsupials like oppossums are found in south America. The marsupials have varied habits and are adapted to various modes of life.

Examples

Thylacinus (Tasmanian wolf) is a **carnivorous marsupial.**

Notoryctus (Marsupial mole) is a burrowing form.

Phascolarctes is an arboreal form.

Myrmecobius is an ant eater.

Parameles is a marsupial bandicoot.

Macropus (*Kangaroo) is the largest marsupial.*

Macropus

Sub-class 3 : EUTHERIA *OR* PLACENTALIA

1. *These are highly evolved mammals with advanced organisation.*
2. *Marsupium and marsupial bones are entirely absent.*
3. Nourishment, respiration and excretion of young take place through a complex *allantoic placenta.*
4. The young ones are born in relatively advanced state (miniature adults). Complete developed takes place in the uterus only. Gestation period is generally long.
5. Testes are usually present in scrotal sacs. Penis is situated infront of the testes.
6. *Vagina is single. Urinogenital passage and anus are separate.*
7. Cerebral hemispheres in brain have more convolutions.
8. *Corpus callosum is well developed.*
9. Eutherians are viviparous.
10. Its distribution is world-wide.

Examples

Rats, bats, rabbits, monkeys, cats, dogs, whales, horses, deers, pigs, elephants, man, etc.

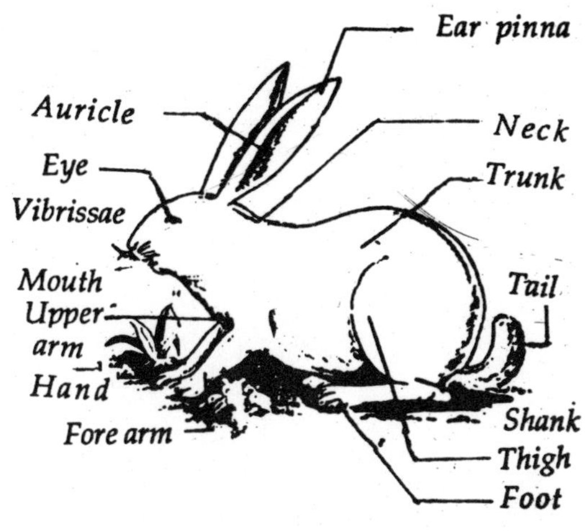

Rabbit

Balaenoptera

Antelope

Rhinoceros